Ordinary differential equations—the building blocks of mathematical modeling—are also key elements of disciplines as diverse as engineering and economics. Although mastery of these equations is essential, adhering to any one method of solving them is not: This book stresses alternative examples and analyses by means of which the student can build an understanding of a number of approaches to finding solutions and understanding their behavior.

The text includes brief expositions of standard topics, including first-order equations, homogeneous and nonhomogeneous second-order linear equations, power series expansions about regular and regular singular points, linear systems theory, and stability concepts for both the phase plane and higher-dimensional systems. A variety of exercises and examples is included, and readers are encouraged to try alternative approaches to find solutions that integrate and build upon ideas introduced in earlier chapters. This book offers not only an applied perspective for the student learning to solve differential equations, but also the challenge to apply these analytical tools in the context of singular perturbations, which arises in many areas of application. An important resource for the advanced undergraduate, this book would be equally useful for the beginning graduate student investigating further approaches to these essential equations.

Thinking About Ordinary Differential Equations

Cambridge Texts in Applied Mathematics

Maximum and Minimum Principles
M.J. Sewell

Solitons
P.G. Drazin and R.S. Johnson

The Kinematics of Mixing
J.M. Ottino

Introduction to Numerical Linear Algebra and Optimisation
Phillippe G. Ciarlet

Integral Equations
David Porter and David S.G. Stirling

Perturbation Methods
E.J. Hinch

The Thermomechanics of Plasticity and Fracture
Gerard A. Maugin

Boundary Integral and Singularity Methods for Linearized Viscous Flow
C. Pozrikidis

Nonlinear Systems
P.G. Drazin

Stability, Instability and Chaos
Paul Glendinning

Applied Analysis of the Navier-Stokes Equations
C.R. Doering and J.D. Gibbon

Viscous Flow
H. Ockendon and J.R. Ockendon

Similarity, Self-similarity and Intermediate Asymptotics
G.I. Barenblatt

A First Course in the Numerical Analysis of Differential Equations
A. Iserles

Complex Variables: Introduction and Applications
Mark J. Ablowitz and Athanssios S. Fokas

Thinking About Ordinary Differential Equations
Robert E. O'Malley, Jr.

Thinking About Ordinary Differential Equations

ROBERT E. O'MALLEY, JR.
University of Washington

PUBLISHED BY THE PRESS SYNDICATE OF THE UNIVERSITY OF CAMBRIDGE
The Pitt Building, Trumpington Street, Cambridge CB2 1RP, United Kingdom

CAMBRIDGE UNIVERSITY PRESS
The Edinburgh Building, Cambridge CB2 2RU, United Kingdom
40 West 20th Street, New York, NY 10011-4211, USA
10 Stamford Road, Oakleigh, Melbourne 3166, Australia

First published 1997

Printed in the United States of America

Typeset in Times Roman

Library of Congress Cataloging-in-Publication Data

O'Malley, Robert E,
Thinking about ordinary differential equations /
Robert E. O'Malley, Jr.
p. cm. – (Cambridge texts in applied mathematics)
Includes bibliographical reference (p. 243) and index.
ISBN 0-521-55314-8 (hardback). – ISBN 0-521-55742-9 (pbk.)
1. Differential equations. I. Title. II. Series.
QA372.057 1997
515'.352-dc20 96-14825
CIP

*A catalogue record for this book is available from
the British Library*

ISBN 0-521-55314-8 hardback
ISBN 0-521-55742-9 paperback

To Candy, Patrick, Timothy, and Daniel . . .
Great O'Malleys, blissfully unenlightened concerning
differential equations

Contents

Preface

This small book is intended for use by students in the applied sciences and engineering who already have some elementary knowledge of ordinary differential equations. It aims to emphasize the variety of analytical approaches available and to teach simple techniques to use in their own technical work and in understanding the behavior of solutions to many standard problems. The exercises at the end of each chapter, in particular, are intended to be the primary learning tool, so fairly detailed solutions are provided for many of them. The important job of interpreting solutions in their underlying physical context is left to the reader.

Good calculus skills are called for. Some familiarity with numerical and/or symbolic computing and with matrix analysis would also be helpful, but is not necessary. We will not hesitate to introduce needed theory, without proof, in order to advance the reader's understanding. The fundamental perspective is that there is no best way to solve a given ordinary differential equation. Indeed, most equations that scientists encounter are solved numerically, and the traditional analytical techniques presented here remain important because they provide the basis for successful computing schemes. Readers are definitely urged to use available software to learn about the solutions of the differential equations they either need to solve or have otherwise become fascinated by.

The examples and exercises included have been collected over many years for various classes given by the author. Many were taken from others' textbooks and papers (only a few are original), so it is no longer possible to properly acknowledge the original sources. This explicit debt to earlier writers is, certainly, substantial. Likewise, little reference is made to more advanced monographs, as would be appropriate for students seeking a less-utilitarian acquintanceship with differential equations.

We hope readers will find the problems considered interesting and challenging. Moreover, we hope they will learn enough about how to solve differential equations to be able to apply the tools introduced in substantial applications. Reading this book is a do-it-yourself project; some scribbling between steps is expected, and fooling around with toy problems is encouraged.

Special thanks are extended to all those who have helped me learn this material and prepare this manuscript. They include Shepley Ross and Bob Owens (years ago), Frances Chen (most recently), and many colleagues and students (over the years).

Robert E. O'Malley, Jr.
March 1996

1

First-order equations

1.1 Introductory remarks

Differential equations are encountered as relations between independent variables, such as t, and an unknown function $x(t)$ of t and some of its derivatives $\frac{d^j x}{dt^j}$, possibly expressed as

$$F\left(t, x, \frac{dx}{dt}, \ldots, \frac{d^n x}{dt^n}\right) = 0 \tag{1.1}$$

on some t interval. We typically seek the unknown $x(t)$ as a scalar, vector, or a matrix, depending on the dimensionality of F.

If partial derivatives with respect to one or more independent variables are involved, we have a *partial differential equation*. We shall, however, primarily consider *ordinary differential equations*, which involve only ordinary derivatives and are frequently expressed as coupled systems

$$\frac{dy_i}{dt} = f_i(y_1(t), y_2(t), \ldots, y_n(t), t), \;\; i = 1, 2, \ldots, n$$

of n scalar equations or, more concisely, as a vector system

$$\frac{dy}{dt} = f(y(t), t) \tag{1.2}$$

for

$$y \equiv \begin{pmatrix} y_1 \\ y_2 \\ \vdots \\ y_n \end{pmatrix}, \;\frac{dy}{dt} = y' \equiv \begin{pmatrix} y_1' \\ y_2' \\ \vdots \\ y_n' \end{pmatrix}, \;\text{and } f \equiv \begin{pmatrix} f_1 \\ f_2 \\ \vdots \\ f_n \end{pmatrix}.$$

More generally, we might encounter a vector system

$$g\left(t, y, \frac{dy}{dt}\right) = 0 \tag{1.3}$$

for, say, n-vectors y, y', and g. When we can solve such a relation to obtain y' as a function of y and t, we say the resulting system $y' = f(y, t)$ is *in standard form*.

Example 1 The scalar differential equation $(y')^2 + y^2 + 1 = 0$ corresponds to two differential equations,

$$y' = \pm i\sqrt{y^2 + 1},$$

in standard form. Complex-valued solutions would result from both possibilities; indeed, the differential equation does not have real solutions. Moreover, the original equation might be allowed solutions that switch (or "chatter") from time to time between the two sign possibilities whenever $y^2 = -1$. We will, however, limit attention below to differential equations with real-valued differentiable solutions.

Example 2 The *differential-algebraic system*

$$\begin{cases} (y_1')^2 = y_1^2 + y_2^2, \\ y_2' = g(y_1, y_2, y_3), \\ y_3^2 + y_2^2 = 4, \end{cases}$$

consisting of two scalar differential equations and one scalar "algebraic" equation, is equivalent to the two-dimensional differential systems

$$\begin{cases} y_1' = \pm\sqrt{y_1^2 + y_2^2}, \\ y_2' = g\left(y_1, y_2, \pm\sqrt{4 - y_2^2}\right) \end{cases}$$

in standard form, whose real solutions are subject to the constraints

$$y_3 = \pm\sqrt{4 - y_2^2} \quad \text{and} \quad -2 \le y_2 \le 2.$$

Any solution

$$\begin{pmatrix} y_1 \\ y_2 \end{pmatrix}$$

of the two-dimensional differential systems satisfying the bound for y_2 defines two possibilities for y_3. As this example suggests, there is plenty of flexibility in defining solutions, although we shall seldom be adventurous.

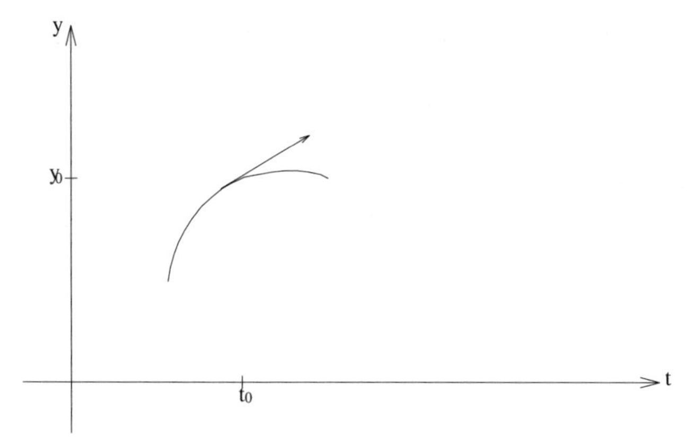

Figure 1. Tangent curve with slope $y'(t_0) = f(y_0, t_0)$ at (y_0, t_0) (scalar case).

If we are given the value $y(t_0)$ at some prescribed initial time t_0, it is natural to seek the solution $y(t)$ of the *initial value problem*

$$\frac{dy}{dt} = f(y(t), t), \quad y(t_0) = y_0 \tag{1.4}$$

for t values near t_0. We can readily approximate the solution as a short curve $y(t)$ through the given point (y_0, t_0) in (y, t) space with an initial *slope* $f(y_0, t_0)$. Note that (1.4) makes sense when y is a scalar, vector, or matrix unknown.

We would expect the solution $y(t)$ of (1.4) to be *unique* and to *exist* as long as $f(y(t), t)$ continues to be smoothly defined (and, in particular, remains bounded). Continuity of f and f_y will, indeed, suffice.

We naturally call the scalar equation

$$F\left(t, x, \frac{dx}{dt}, \ldots, \frac{d^n x}{dt^n}\right) = 0 \tag{1.1}$$

(for an unknown $x(t)$) an *nth order* differential equation, since the highest derivative involved is the nth (presuming actual dependence on the last argument of F). It can always be converted to an equivalent system of n first-order equations by, for example, setting

$$\begin{cases} y_1 = x \\ y_2 = \frac{dy_1}{dt} = x' \\ \vdots \\ y_{n-1} = \frac{dy_{n-2}}{dt} = x^{(n-2)} \\ y_n = \frac{dy_{n-1}}{dt} = x^{(n-1)}. \end{cases} \tag{1.5}$$

Since $\frac{dy_n}{dt} = x^{(n)}$, we can rewrite the original differential equation (1.1) as

$$F\left(t, y_1, y_2, \ldots, y_n, \frac{dy_n}{dt}\right) = 0.$$

Thus, a first-order vector system, equivalent to the nth-order scalar equation (1.1), for the vector variable y defined componentwise by (1.5), is

$$\begin{cases} \frac{dy_1}{dt} = y_2 \\ \frac{dy_2}{dt} = y_3 \\ \vdots \\ \frac{dy_{n-1}}{dt} = y_n \\ F\left(t, y_1, y_2, \ldots, y_n, \frac{dy_n}{dt}\right) = 0. \end{cases} \tag{1.6}$$

This system has a very special structure, suggesting that the general *first-order vector system*

$$g\left(t, y, \frac{dy}{dt}\right) = 0 \tag{1.7}$$

(for an arbitrary n vector g) is considerably more inclusive than higher-order scalar equations (1.1). Scalar equations, especially those of second order, nonetheless remain basic because they, for example, include Newton's law $\mathcal{F} = m\frac{d^2x}{dt^2}$ and many other fundamental laws of physics and engineering. Readers should become comfortable working with both first- and higher-order systems of ordinary differential equations.

Let us continue by considering two first-order examples from *population dynamics*.

Example 3 Suppose the population $N(t)$ of a given species (bacteria or elves) is not always zero and varies at a rate proportional to its current value, i.e.

$$\frac{dN}{dt} = rN \tag{1.8}$$

where r is some measured constant proportionality factor. Suppose the initial population $N(0) > 0$ is given. [If $N(0) = 0$, $\frac{dN}{dt}(0) = 0$ implies that $N(t) \equiv 0$ for all $t \geq 0$. (Without Adam and Eve, none of us would be here.)] Indeed, $N(t) \neq 0$ for all finite t (since $N(t_0) = 0$ and the change of independent variable $s = t - t_0$ yields $\frac{dN}{ds} = rN$ with $N = 0$ at $s = 0$.) Integrating $\frac{dN}{N} = r\,dt$ (since $N \neq 0$) implies that $\ln\left(\frac{N(t)}{N(0)}\right) = rt$, so

$$N(t) = e^{rt}N(0). \tag{1.9}$$

Thus,

$N(t)$ remains constant if $r = 0$;

$N(t)$ increases exponentially with t if $r > 0$;

and

$N(t)$ decreases monotonically, tending to zero as $t \to \infty$, if $r < 0$.

(Demographers: Is it realistic or relevant to let t become unbounded?)

Example 4 Worrying about a limited food supply and a constantly growing population, we might replace any positive proportionality constant r in the preceding *Malthusian* model (1.8) by a more biologically motivated *logistic function* $f(N)$, which will ultimately become negative for large values of N. Taking $f(N) = r\left(1 - \frac{N}{K}\right)$, for simplicity, we obtain

$$\frac{dN}{dt} = r\left(1 - \frac{N}{K}\right)N \tag{1.10}$$

for a constant saturation level (or *carrying capacity*) $K > 0$ and an *intrinsic growth rate* $r > 0$. (Our preceding model thereby corresponds to the limiting possibility $K = \infty$.) Quadratic differential equations like (1.10) are usually called *Riccati equations*. They will be considered further below, because, although they are not simple, they are explicitly solvable.

We are able to determine the qualitative behavior of the population $N(t)$ as t varies without explicitly solving the differential equation. Simply note that

$$\frac{dN}{dt} > 0 \text{ for } 0 < N < K,$$

whereas

$$\frac{dN}{dt} < 0 \text{ for } K < N$$

and N remains constant at $N = 0$ and $N = K$. This implies that

$$\begin{cases} N(t) \equiv 0 & \text{if } N(0) = 0, \\ N(t) \uparrow K & \text{if } 0 < N(0) < K, \text{ where the } \textit{steady-state } K \text{ is} \\ & \text{attained from below as } t \to \infty, \\ N(t) \equiv K & \text{if } N(0) = K, \\ \text{and} & \\ N(t) \downarrow K & \text{if } N(0) > K \end{cases} \tag{1.11}$$

Thus, the solution $N(t)$ of the initial value problem is always a monotonic function of the independent variable $t \geq 0$, and the limit K is attained for $N(t)$ as

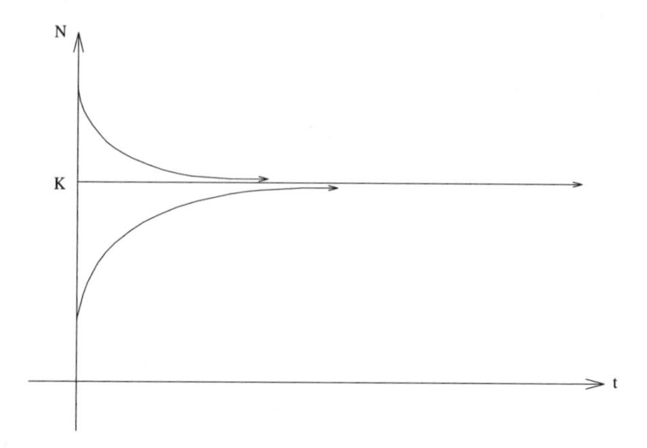

Figure 2. Population decay to carrying capacity.

$t \to \infty$, unless $N(0) = 0$. Pictorially, we have the situation qualitatively depicted in Figure 2. (The arrow indicates the direction of motion as t increases.) We naturally say that all positive initial values $N(0)$ lie in the *domain of attraction* of the stable *steady state* or equilibrium K (which is only attained as t becomes unbounded, unless $N(t) \equiv K$).

Readers are urged to investigate solutions of differential equations like (1.10) for a variety of initial conditions using readily available software such as Matlab or Mathematica to obtain and graph solutions (see, e.g., Polking (1995)). Much insight into the behavior of solutions can be gained by such experimentation, a possibility not available to those first studying these equations in the eighteenth century or to those who modeled various populations in the early twentieth century.

We can explicitly find the solution $N(t)$ of (1.10) by *separating variables*:

$$\frac{K dN}{(K - N)N} = r dt.$$

(This does not require dividing by zero, since only the constant populations $N(t)$ can ever equal 0 or K.) Using *partial fractions* (a technique that will be of recurring value below), we rewrite this equation in terms of constants A and B,

$$\frac{K}{(K - N)N} = \frac{A}{N} + \frac{B}{K - N},$$

where A and B which must satisfy

$$K = A(K - N) + BN = AK + (B - A)N.$$

Equating coefficients of both powers of N (i.e., 0 and 1) to achieve a balance for small and large N values, we ask that $K = AK$ and $B - A = 0$, yielding $B = A = 1$. Thus, the separated differential equation implies that

$$\frac{dN}{N} + \frac{dN}{K - N} = r dt.$$

If we integrate both sides, this implies

$$\ln|N| - \ln|K - N| = rt + C$$

for some constant C, so

$$\ln\left|\frac{N(t)}{K - N(t)}\right| = rt + C.$$

Exponentiating, we get $\frac{N(t)}{K-N(t)} = \widetilde{C}e^{rt}$, where the still arbitrary constant $\widetilde{C} = \pm e^{C}$ now becomes specified through the initial condition as $\widetilde{C} = \frac{N(0)}{K-N(0)}$. Solving for $N(t)$, we finally obtain the explicit solution

$$N(t) = \frac{KN(0)}{N(0) + (K - N(0))e^{-rt}}, \quad t \geq 0 \tag{1.12}$$

to example 4, which remains valid as t increases as long as the denominator stays nonzero. The validity of (1.12) can be directly checked by substitution in the differential equation (1.10). Note that the constant solutions are obtained for $N(0) = 0$ and for $N(0) = K$. For $0 < N(0) < K$, both terms in the denominator remain positive, but the second decreases monotonically from $K - N(0)$ to 0 as t increases; thus, $N(t) \uparrow K^{-}$ as $t \to \infty$. If $N(0) > K > 0$, the denominator $Ke^{-rt} + N(0)(1 - e^{-rt})$ also remains positive for all finite t, decreasing from K to $N(0)$ as t increases, so the steady-state population K is then monotonically approached from above as $t \to \infty$. Anticipating later terminology, note that the "nonlinear" population model (1.10) is somewhat more complicated to explicitly solve than the more elementary "linear" model (1.8). The saturation obtained from the nonlinear model at the finite carrying capacity K should provide a more realistic conclusion, however, than continuing exponential growth. One must check the numbers obtained against reality (before calling the exterminator).

Formula (1.12) provides the solution of the Riccati equation (1.10) as long as it exists for any set of *parameters* K, r, and $N(0)$. If we allow $r < 0$, one can readily show that $N(t) \downarrow 0$ as t increases if $0 < N(0) < K$, whereas

$N(t) \uparrow \infty$ as t increases for $N(0) > K$. The constant solution $N(t) \equiv K$ for $N(0) = K$ is naturally called *unstable* to perturbations of the initial value, because a small change in $N(0)$ away from K ultimately results in a far different solution value $N(t)$; the solution $N(t) \equiv 0$, by contrast, is a *stable* constant solution. An equivalent conclusion would be reached for $r > 0$, if we instead let t decrease from zero, but then we would be determining the population of ancestors rather than descendents, as geneologists might prefer. The important concepts of stability and instability will be the practical and significant focus of some later effort.

Generally, an nth-order scalar differential equation $F\left(t, x, \frac{dx}{dt}, \dots, \frac{d^n x}{dt^n}\right) = 0$ is called *linear* if it can be written in the very restricted form

$$a_n(t)\frac{d^n x}{dt^n} + a_{n-1}(t)\frac{d^{n-1}x}{dt^{n-1}} + \dots + a_1(t)\frac{dx}{dt} + a_0(t)x = f(t) \tag{1.13}$$

for coefficients $a_n, a_{n-1}, \dots, a_0$, and an f that depends only on t, with the leading coefficient $a_n(t)$ being nonzero somewhere. (Note that equation (1.13) involves, in particular, no products of x and/or its derivatives.) All other nth-order scalar differential equations are called *nonlinear* (i.e., not linear). Furthermore, a linear equation (1.13) is called *homogeneous*, and it has at least the *trivial solution* $x(t) \equiv 0$, if $f(t) \equiv 0$. Otherwise, it is called *nonhomogeneous*, and doesn't have the trivial solution. Note that arbitrary *linear combinations* $\alpha x(t) + \beta z(t)$ of any solutions $x(t)$ and $z(t)$ of such a linear homogeneous equation also satisfy the equation, whatever the constants α and β. This *superposition principle* is a cornerstone of many methods to solve all kinds of linear equations.

Analogously, a first-order differential system $g\left(t, y, \frac{dy}{dt}\right) = 0$ of n scalar equations for an n-vector $y(t)$ is called linear if it can be rewritten in the special vector-matrix form

$$B(t)\frac{dy}{dt} = C(t)y + k(t). \tag{1.14}$$

Here,

$$y = \begin{pmatrix} y_1 \\ \vdots \\ y_n \end{pmatrix}, k(t) = \begin{pmatrix} k_1(t) \\ \vdots \\ k_n(t) \end{pmatrix}, \quad \text{and} \quad \frac{dy}{dt} = \begin{pmatrix} \frac{dy_1}{dt} \\ \vdots \\ \frac{dy_n}{dt} \end{pmatrix}$$

are all n-vectors, while

$$B(t) = \begin{pmatrix} b_{11}(t) & \dots & b_{1n}(t) \\ \vdots & & \\ b_{n1}(t) & \dots & b_{nn}(t) \end{pmatrix}$$

and $C(t)$ are $n \times n$ matrices. We will assume that B, C and k have continuous scalar entries. At *ordinary* points, where $B(t)$ is nonsingular, we can multiply (1.14) by the inverse $B^{-1}(t)$ to get a linear differential system

$$\frac{dy}{dt} = A(t)y + g(t) \tag{1.15}$$

in standard form (with $A \equiv B^{-1}C$ and $g \equiv B^{-1}k$). Alternatively, the unknown solution components $y_i(t)$ must then successively satisfy the corresponding n scalar coupled linear differential equations

$$\frac{dy_i}{dt} = a_{i1}(t)y_1(t) + \ldots + a_{in}(t)y_n(t) + g_i(t), \quad i = 1, 2, \ldots, n.$$

The linear system (1.15) is homogeneous, and has the trivial solution $y(t) \equiv 0$, when $g(t) \equiv 0$. In the *autonomous* situation when all n^2 entries of the matrix $A(t) \equiv A$ are constant, it will eventually be convenient to define the matrix exponential e^{At} as a matrix solution Y of the constant linear homogeneous system $Y' = AY$. As before, nonlinear, nonhomogeneous, and nonautonomous, respectively, simply mean not linear, not homogeneous, and not autonomous. Linear differential systems (1.14) will be shown to be quite tractable, except near values of t where the matrix $B(t)$ becomes singular (i.e., *singular* points). This is why knowing how to solve such linear problems has been critical to advancing our knowledge of physics.

Example 5 Now consider the nonlinear second-order scalar differential equation

$$\frac{d^2r}{dt^2} = -g\frac{R^2}{r^2}, \tag{1.16}$$

which describes the motion of a *free-falling body* of unit mass toward the Earth, where $g > 0$ is the acceleration of gravity, $r \geq R > 0$ is the distance from the center of the Earth, and R is the Earth's radius. (When $r \approx R$, one usually approximates the right-hand side by $-g$ and calculus provides the quadratic solution $r(t) = -\frac{g}{2}t^2 + r'(0)t + r(0)$ of the resulting linear model.) If we (cleverly) multiply equation (1.16) by $2\frac{dr}{dt}$ (which is nonzero when the body is still falling), we get $2\frac{dr}{dt}\frac{d^2r}{dt^2} = -2gR^2\frac{1}{r^2}\frac{dr}{dt}$ or

$$\frac{d}{dt}\left(\left(\frac{dr}{dt}\right)^2 - 2gR^2\frac{1}{r}\right) = 0.$$

Integrating once, we then obtain

$$\left(\frac{dr}{dt}\right)^2 = \frac{2gR^2}{r} + C,$$

which can be interpreted as a *conservation of energy* statement, where the constant $C = (r'(0))^2 - \frac{2gR^2}{r(0)}$ is the sum of a *kinetic energy* $\left(\frac{dr}{dt}\right)^2$ and a *potential energy* $-\frac{2gR^2}{r}$ (which terms separately vary with time). Observe that C is completely determined by prescribing the initial position $r(0)$ and initial velocity $r'(0)$. Solving for

$$\frac{dr}{dt} = -\sqrt{\frac{2gR^2}{r} + C},$$

with the negative velocity selected because the fall is toward the Earth, we obtain the separable first-order equation

$$\frac{dr}{dt} = -\sqrt{(r'(0))^2 + 2gR^2\left(\frac{1}{r} - \frac{1}{r(0)}\right)}$$

for $r(t)$. The implicit (decreasing) solution $r(t)$ is thereby uniquely given by

$$t = -\int_{r(0)}^{r} \frac{ds}{\sqrt{(r'(0))^2 + 2gR^2\left(\frac{1}{s} - \frac{1}{r(0)}\right)}}, \tag{1.17}$$

since $r = r(0)$ at $t = 0$. The explicit solution $r(t)$ might be determined numerically or by inverting the graph of t as a function of r. When $r = R$, the body collides with the Earth and we must certainly question the continued appropriateness of this differential equation model.

Example 6 A simple nonlinear system of differential equations expresses the *law of mass action* in chemical kinetics. Consider a reaction described symbolically by

$$A + B \xrightarrow{k} C,$$

where $k > 0$ is a given *rate constant*. Letting u, v, and w denote the (nonnegative) concentrations of the chemicals A, B, and C, respectively, we obtain a system of three coupled nonlinear first-order differential equations

$$\begin{cases} \frac{du}{dt} = -kuv \\ \frac{dv}{dt} = -kuv \\ \frac{dw}{dt} = kuv, \end{cases} \tag{1.18}$$

assuming that the rates of change of all of the concentrations are equal to the product of the concentrations of the reactants A and B times the rate constant. If $u(0) > 0$ and $v(0) > 0$ are prescribed and $w(0) = 0$, we will have $\frac{d}{dt}(u + w)$

$= \frac{d}{dt}(u - v) = 0$, so $u + w = u(0)$ and $u - v = u(0) - v(0)$ allows us to eliminate

$$w(t) = u(0) - u(t) \quad \text{and} \quad v(t) = u(t) + v(0) - u(0), \tag{1.19}$$

reducing our three-component problem (1.18) to only the single separable differential equation

$$\frac{du}{dt} = -ku[(u + v(0) - u(0))] \tag{1.20}$$

for the single concentration u on $t \geq 0$. Equation (1.20) is a Riccati equation, like that already encountered in the nonlinear population model (1.10), so we need not consider its solution any further (cf. (1.12)). Note that the relations (1.19) describe natural chemical balances. Further, (1.20) shows that u will tend monotonically to $u(0) - v(0)$ as t increases if this value is positive and to 0 otherwise.

A general, but not directly helpful, result is that the (vector or scalar) initial value problem

$$\frac{dy}{dx} = f(x, y), \quad y(x_0) = y_0 \tag{1.21}$$

has a unique solution

$$y = \begin{pmatrix} y_1 \\ \vdots \\ y_n \end{pmatrix}$$

in some interval $x_0 - h < x < x_0 + h$ for a (possibly small) $h > 0$, provided the n-vector

$$f = \begin{pmatrix} f_1 \\ \vdots \\ f_n \end{pmatrix}$$

and the $n \times n$ *Jacobian* matrix

$$\frac{\partial f}{\partial y} = \begin{pmatrix} \frac{\partial f_1}{\partial y_1} & \cdots & \frac{\partial f_1}{\partial y_n} \\ \vdots & & \\ \frac{\partial f_n}{\partial y_1} & \cdots & \frac{\partial f_n}{\partial y_n} \end{pmatrix}$$

have continuous entries in an $(n + 1)$-dimensional neighborhood of the prescribed (bounded) initial point (x_0, y_0). This is the basic *existence-uniqueness theorem* for smooth ordinary differential equations in standard form. Its consequences will include many practical results. We shall, in particular, make extensive use of the conclusion. Note that weaker hypotheses which separately guarantee existence and uniqueness are also available.

Integrating (21) from x_0 shows that the initial value problem is equivalent to the *integral equation*

$$y(x) = y_0 + \int_{x_0}^{x} f(s, y(s))ds \tag{1.22}$$

for the unknown $y(x)$. (Any differentiable solution of this integral equation will satisfy the initial value problem and vice versa.) Following Picard (in about 1890), we might try to solve the integral equation using *successive approximations*, by defining

$$y^0(x) \equiv y_0,$$

$$y^1(x) \equiv y_0 + \int_{x_0}^{x} f(s, y_0)ds,$$

$$y^2(x) \equiv y_0 + \int_{x_0}^{x} f\left(s, y_0 + \int_{x_0}^{s} f(r, y_0)dr\right) ds,$$

and, in general, for any $m \geq 1$,

$$y^{m+1}(x) \equiv y_0 + \int_{x_0}^{x} f(s, y^m(s))ds. \tag{1.22}$$

The iterates y^m actually converge pointwise near x_0 (though, perhaps, very slowly) to a unique solution

$$y(x) \equiv \lim_{m\to\infty} y^m(x) \tag{1.23}$$

under the stated smoothness conditions. Not surprisingly, the *iteration procedure* also provides the basis for a numerical (finite difference) method. We could even be more flexible in selecting the initial iterate y^0. Furthermore, for any linear system $y' = A(x)y + b(x)$, with $y(x_0) = y_0$ prescribed and $A(x)$ and $b(x)$ everywhere continuous, the solution obtained actually exists for *all* finite x.

Complete proofs of these statements are more advanced than the intended level of this textbook, but our coverage below should convince the reader of the theorem's validity and importance. We will now discuss analytical solution methods for several special classes of first-order scalar equations.

1.2 Separable equations

Consider the (somewhat general) equation

$$h(y)\frac{dy}{dx} = g(x); \tag{1.24}$$

we separate variables to get

$$h(y)dy = g(x)dx.$$

(Realize, however, that you may encounter singularities of h or g, making it advisable to check any answer obtained.) Next, integrate both sides of this equation with respect to their separate variables to obtain the relation

$$\int^y h(r)dr = \int^x g(s)ds \tag{1.25}$$

for $y(x)$. If we impose the initial condition $y(x_0) = y_0$, the solution $y(x)$ or $x(y)$ will be implicitly defined by the integrals

$$\int_{y_0}^y h(r)dr = \int_{x_0}^x g(s)ds,$$

presuming they both remain defined as the path of integration moves from the initial point (x_0, y_0) to a general point (x, y). In the special case when h is constant, we get simply the usual problem of integral calculus. A differential equation can sometimes be put in the special form (1.24) through a preliminary transformation. For example, $z' = xe^{x+z} - 1$ is separated by introducing $y = x + z$, while $xu' = uF(xu)$ takes the form (1.24) after we set $y = xu$.

Example 7 The initial value problem

$$\frac{dy}{dx} = \frac{y\cos x}{1+2y^2}, \quad y(0) = 1$$

is separable, allowing us to rewrite the equation as

$$\left(\frac{1+2y^2}{y}\right)dy = \frac{dy}{y} + 2ydy = \cos x dx$$

for $y \neq 0$. (Beware that we can lose solutions, like $y(x) \equiv 0$ in this example, when we separate variables.) Integrating both sides, we get

$$\ln|y(x)| + y^2(x) = \sin x + C$$

for some constant C. Thus $y(0) = 1$ implies that $C = 1$. Any solution $y(x)$ is therefore implicitly defined by

$$H(y^2) \equiv \tfrac{1}{2}\ln(y^2(x)) + y^2(x) = 1 + \sin x.$$

Note that $y(x) \neq 0$ or ∞ since $1 + \sin x$ remains bounded between 0 and 2 for all x. Note further that the positive solution y is locally unique, by the *implicit function theorem*, because $\frac{dH}{dy^2} = \frac{1}{2y^2} + 1 > 0$ implies that y^2 is uniquely determined for each x. Also, since $1 + \sin x$ is 2π-periodic, so is the solution $y(x)$. Moreover, the differential equation shows that $\frac{dy}{dx}$ has the same sign as $\cos x$. This implies that the critical (extreme) values $y_{\max,\min}$ will be attained when $x = \frac{\pi}{2}$ and $\frac{3\pi}{2}$, respectively, so $y^2_{\max} + \frac{1}{2}\ln\left(y^2_{\max}\right) = 2$ and $y^2_{\min} + \frac{1}{2}\ln\left(y^2_{\min}\right) = 0$, i.e. $y_{\min} \approx 0.653$ and $y_{\max} \approx 1.32$. These values could be verified by using a graphing calculator or a numerical method to directly approximate the solution to the differential equation or its implicit representation. We would then obtain

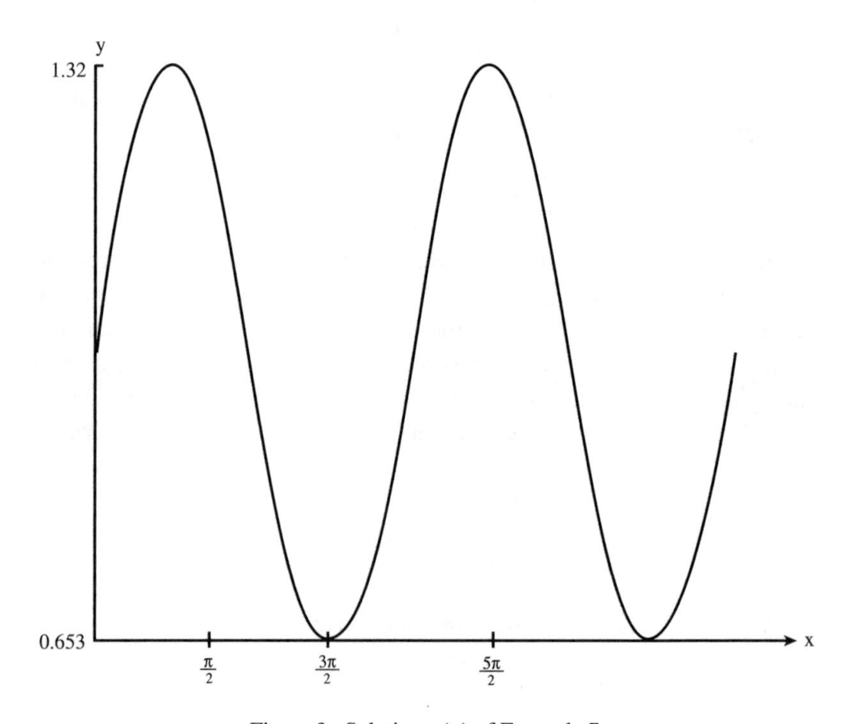

Figure 3. Solution $y(x)$ of Example 7.

Because $y(x + 2k\pi) = y(x)$ for all integers k, we need only find the solution in the interval $0 \leq x < 2\pi$ to define it for all x.

Example 8 Suppose

$$\frac{dy}{dx} = x^2 y^2.$$

For $y \neq 0$, we have

$$\frac{dy}{y^2} = x^2 dx \text{ or } d\left(-\frac{1}{y}\right) = d\left(\frac{x^3}{3}\right),$$

so integration implies $-\frac{1}{y} = \frac{x^3}{3} + C$ and $y = \frac{-3}{x^3+3C}$, presuming the denominator remains nonzero. If $y(0) = 0$, we get the trivial solution $y(x) \equiv 0$. Otherwise, we solve for $C = -1/y(0)$ to get the unique solution

$$y(x) = \frac{3y(0)}{3 - y(0)x^3},$$

which remains defined as long as $y(0)x^3 \neq 3$. If $y(0) > 0$, the solution is defined for all $x < \sqrt[3]{3/y(0)}$; whereas if $y(0) < 0$, the solution *blows up* as $x \to -\sqrt[3]{-3/y(0)}$ from above. One says the solution y exhibits a *finite escape* at these bounded limiting x values.

The function

$$y(x) = \frac{3y(0)}{3 - y(0)x^3}$$

is certainly defined for large values of x, but it is critical that readers note that there is *no* solution $y(x)$ of the differential equation, parameterized by the independent variable x, joining $y(0) > 0$ and the same function $y(x)$ for large x values. Blowup occurs in between. Also note that the (sometimes-called) *general solution*

$$y(x) = \frac{-3}{x^3 + 3C}$$

of this nonlinear differential equation, with arbitrary C, does not contain the *trivial solution* $y(x) \equiv 0$ as a special case, unless we allow the limiting value $C = \infty$. Note that the solution's interval of existence would differ if the initial value $y(x_0)$ were prescribed at some point $x_0 \neq 0$.

Example 9 The linear homogeneous equation

$$\frac{dy}{dx} + a(x)y = 0,$$

with a continuous coefficient $a(x)$, is of fundamental importance. It can be solved as the separated equation

$$\frac{dy}{y} = -a(x)dx,$$

if we presume $y \neq 0$. This implies the solution $\ln|y(x)| = -\int^x a(s)ds + C$, which can be more conveniently expressed as

$$y(x) = e^{-\int^x a(s)ds} K$$

for $K = \pm e^C$ as an arbitrary constant. If $y(x_0)$ is prescribed, $y(x_0) = \exp[-\int^{x_0} a(s)ds]K$ determines K and implies that the unique solution of the corresponding initial value problem is

$$y(x) = e^{-\int_{x_0}^x a(s)ds} y(x_0).$$

This answer can be checked (for all x) by direct substitution into both the given differential equation and the initial condition. This solution could only cease to exist if $a(x)$ lost its smoothness or if $|x|$ became infinite. Note that the formula even provides the unique trivial solution when $y(x_0) = 0$.

1.3 Exact equations

The solution $y(x)$ of the linear equation $y' + a(x)y = 0$ satisfies $\frac{d}{dx}[\exp(\int^x a(s)ds)y(x)] = 0$ since differentiation of this product requires that

$$e^{\int^x a(s)ds}(y' + a(x)y) = 0.$$

Consequently, multiplying the given linear homogeneous equation through by the nonzero *integrating factor* $\pm\exp[\int^x a(s)ds]$ makes it *exact*, so $\exp[\int^x a(s)ds]y(x)$ remains constant as x varies, allowing us to solve directly for $y(x)$. We can always solve linear equations by converting them to so-called *exact* equations.

Example 10 Consider

$$2xy^3 + 3x^2y^2\frac{dy}{dx} = 0.$$

The equation is linear and homogeneous, since it is equivalent to $\frac{dy}{dx} = -\frac{2y}{3x}$ provided $xy \neq 0$. However, the equation can also be immediately integrated as the exact equation

$$\frac{d}{dx}(x^2y^3) = 0.$$

Thus, $x^2y^3 = C^3$ along solutions, for some constant C^3, and it follows that

$$y(x) = x^{-2/3}C \quad \text{for } x \neq 0,$$

with the constant C to be fixed by prescribing an initial condition at any $x_0 \neq 0$. (Alternatively, we could multiply the equivalent linear equation $\frac{dy}{dx} + \frac{2}{3x}y = 0$ through by the integrating factor

$$e^{\frac{2}{3}\int^x \frac{ds}{s}} = e^{\frac{2}{3}\ln|x|+k} \equiv |x|^{2/3}\tilde{c} = \pm x^{2/3}\tilde{c}$$

for $x \neq 0$ to get $\frac{d}{dx}\left(x^{2/3}y\right) = x^{2/3}y' + \frac{2}{3}\frac{y}{x^{1/3}} = 0$, again yielding $x^{2/3}y = C$ for $x \neq 0$.)

More generally, suppose we can rewrite a first-order differential equation in the form

$$M(x, y)dx + N(x, y)dy = 0. \tag{1.26}$$

We will call equation (1.26) *exact* if it has the (differential) form

$$d(\psi(x, y)) = 0$$

for some smooth function $\psi(x, y)$. Because the chain rule implies that $d\psi = \psi_x dx + \psi_y dy = 0$, (1.26) essentially requires us to solve the coupled pair of partial differential equations

$$\psi_x = M \quad \text{and} \quad \psi_y = N$$

for a constant function ψ. In order that the mixed derivatives ψ_{xy} and ψ_{yx} to be equal, we therefore also need to satisfy the *exactness condition*

$$M_y = N_x. \tag{1.27}$$

Matters are getting complicated! Being able to satisfy (1.27) is a lot to require for an arbitrary equation (1.26). We are more apt to be successful in satisfying such a condition if we first multiply the equation through by an appropriate (nonzero) integrating factor $\mu(x, y)$ that converts (1.26) to a new equivalent equation

$$\mu(x, y)M(x, y)dx + \mu(x, y)N(x, y)dy = 0$$

satisfying the exactness condition

$$(\mu(x, y)M(x, y))_y = (\mu(x, y)N(x, y))_x. \tag{1.28}$$

Then, we seek an implicit solution $\varphi(x, y) = c$ that satisfies

$$\varphi_x(x, y) = \mu(x, y)M(x, y),$$

which implies that

$$\varphi(x, y) = \int^x \mu(s, y)M(s, y)ds + K(y) = c \tag{1.29}$$

for some "constant" of integration $K(y)$. Differentiating under the integral sign, and using (1.28), shows that

$$\varphi_y = \int^x \frac{\partial}{\partial y}(\mu(s, y)M(s, y))ds + K'(y) = \mu(x, y)N(x, y).$$

We can also think of (1.28) as a partial differential equation for μ. One can anticipate making the solution of (1.26) more manageable by looking for integrating factors μ that have some special form, for instance, being functions of x or y only or products of powers of them.

Substantial luck (insight?) is often involved in finding a μ which converts a given differential equation into an exact one. We won't try to become experts at this. Instead, we will remain curious amateurs and simply illustrate the elusive approach through examples. Note that the separated equation $g(x)dx - h(y)dy = 0$ satisfies the exactness condition, so it has the implicit solution

$$\psi(x, y) = \int^x g(s)ds - \int^y h(t)dt.$$

Example 11 The equation

$$\frac{dy}{dx} = -\frac{y(y + 3x)}{x(x + y)} \quad \text{or} \quad (3xy + y^2)dx + (x^2 + xy)dy = 0$$

is not exact, since $(3xy + y^2)_y \neq (x^2 + xy)_x$ shows that the exactness condition (1.27) does not hold. Multiplying by the integrating factor $\mu \equiv x \neq 0$, however, yields the exact equation

$$(3x^2y + xy^2)dx + (x^3 + x^2y)dy = 0. \tag{$*$}$$

Because this equation is equivalent to $d(x^3y + \frac{1}{2}x^2y^2) = 0$,

$$x^3y + \frac{1}{2}x^2y^2 = \frac{c}{2}$$

is constant along solutions. Thus, $(y + x)^2 = x^2 + c/x^2$ provides two real solutions

$$y = -x \pm \sqrt{x^2 + \frac{c}{x^2}},$$

as long as $x^2 + \frac{c}{x^2} \geq 0$ and $x \neq 0$. If we are given an initial value $y(x_0)$ at some $x_0 \neq 0$, say $y(1) = 5$, we determine the unique solution

$$y(x) = -x + \sqrt{x^2 + \frac{35}{x^2}}$$

for all $x > 0$. For $y(1) = -5$, we would instead have

$$y(x) = -x - \sqrt{x^2 + \frac{15}{x^2}}$$

for $x > 0$. The initial value $y(x_0)$ determines which sign to use as well as the constant c.

More mechanically, if we seek a conserved quantity $\varphi(x, y)$ for the exact equation $(*)$ such that

$$\varphi_x = 3x^2y + xy^2 \quad \text{and} \quad \varphi_y = x^3 + x^2y,$$

we could simply integrate the first partial differential equation to obtain

$$\varphi(x, y) = x^3y + \frac{x^2y^2}{2} + h(y)$$

for some unspecified function h. Then, $\varphi_y = x^3 + x^2y + h'(y)$ implies that h must be constant, so we again find that $x^3y + x^2y^2/2 = c/2$. Observe that the tough part of the solution process was guessing the integrating factor x. For this problem, another integrating factor is (purportedly) $\mu(x, y) = [xy(y + 2x)]^{-1}$. Check it!

Example 12 Let us now find the most general function $N(x, y)$ such that the equation

$$(x^3 + xy^2)dx + N(x, y)dy = 0$$

is exact. Our exactness condition, $N_x = 2xy$, requires $N(x, y) = x^2y + g(y)$ for some unspecified g. The differential equation

$$d\psi = \psi_x dx + \psi_y dy = (x^3 + xy^2)dx + (xy^2 + g(y)) = 0$$

is therefore equivalent to the conservation statement that

$$\psi(x, y) = \frac{x^4}{4} + \frac{x^2y^2}{2} + \int^y g(s)ds = 0,$$

since $d\psi \equiv 0$ specifies ψ up to the arbitrary constant anticipated by the indefinite integral. Seeking $x(y)$, we will have

$$(x^2 + y^2)^2 = y^4 - 4\int^y g(s)ds,$$

presuming $y^4 \geq 4\int^y g(s)ds$. Then, however,

$$x^2 = -y^2 + \sqrt{y^4 - 4\int^y g(s)ds} \geq 0$$

(since we must reject the minus sign possibility to keep x real, provided $\int^y g(s)ds \leq 0$). Under this assumption, we obtain the two real solutions

$$x(y) = \pm\sqrt{-y^2 + \sqrt{y^4 - 4\int^y g(s)ds}}.$$

We cannot expect a unique solution without specifying an appropriate initial condition.

1.4 Linear equations

The integrating factor idea, introduced for exact equations, provides us a direct way to solve all scalar linear differential equations

$$\frac{dy}{dx} + a(x)y = f(x) \tag{1.30}$$

on x intervals where a and f are both given continuous functions. (Note carefully that equation (1.30) has been written in standard form with y' having the coefficient 1 and the linear term ay placed to the left of the equality sign.) If we multiply (1.30) by the nonzero factor $\exp[\int_b^x a(s)ds]$ for some convenient constant b, we get the exact equation

$$\frac{d}{dx}\left(e^{\int_b^x a(s)ds}y(x)\right) = e^{\int_b^x a(s)ds}(y' + a(x)y) = e^{\int_b^x a(s)ds}f(x).$$

Integrating, then, we have

$$e^{\int_b^x a(s)ds}y(x) = \int^x e^{\int_b^t a(s)ds}f(t)dt + C,$$

for some constant C (already implied by the unspecified lower limit of the t integration), so dividing by the exponential integrating factor, we get the general solution

$$y(x) = \int^x e^{-\int_t^x a(s)ds}f(t)dt + e^{-\int^x a(s)ds}C. \tag{1.31}$$

(Note our careful attempt to distinguish between independent variables outside integrals and dummy variables of integration within them.) You can directly check that the first term on the right-hand side of (1.31) is a (so-called) particular solution of the linear nonhomogeneous equation, whereas the second term (the so-called complementary solution) is the general solution of the corresponding homogeneous equation since C is arbitrary. Any other particular solution could be substituted for the first term. Indeed, the second term is redundant because the lower limit of integration in the first term is unspecified. In this way, we see that the linear vector space of solutions for the linear differential equation (1.30) mimics the affine structure for solutions of the linear algebraic system $Ax = b$. (More linear algebra will be fundamental in Chapter 4, where linear matrix differential systems are studied, so it's time to learn or brush up on your matrix theory!)

More simply, observe that an alternative derivation of (1.31) can be obtained by seeking a solution of (1.30) in the so-called *variation-of-parameters* form

$$y(x) = e^{-\int^x a(s)ds} z(x).$$

where any constant multiple of the factor $\exp[-\int^x a(s)ds]$ satisfies the homogeneous equation. Substitution of y into (1.30) require the unknown z to satisfy $z' = \exp[\int^x a(s)ds]f(x)$, so

$$z(x) = \int^x e^{\int^t a(s)ds} f(t)dt.$$

Let us now impose the initial value $y(x_0)$, so C becomes specified in (1.31) and the linear initial value problem for (1.30) is shown to have the *unique solution*

$$y(x) = \int_{x_0}^x e^{-\int_t^x a(s)ds} f(t)dt + e^{-\int_{x_0}^x a(s)ds} y(x_0). \tag{1.32}$$

If either a and f were discontinuous, then this formula could only be used as long as the integrals remained defined. In particular, notice that the existence-uniqueness theorem does not apply to the equation $xy' = y + x^2 \cos x$, for the initial point $x = 0$ (cf. also Exercise 27).

We apologize that the exponential expressions in (1.32) are bulky. It is sometimes convenient to set

$$\int_t^x a(s)ds \equiv (x - t)\bar{a}(x, t),$$

recognizing that $\bar{a}(x, t)$ is an average of the coefficient $a(s)$ over the integration interval.

Example 13 Consider the linear initial value problem

$$\frac{dy}{dx} + \frac{y}{x} = 1, \quad y(1) = 0$$

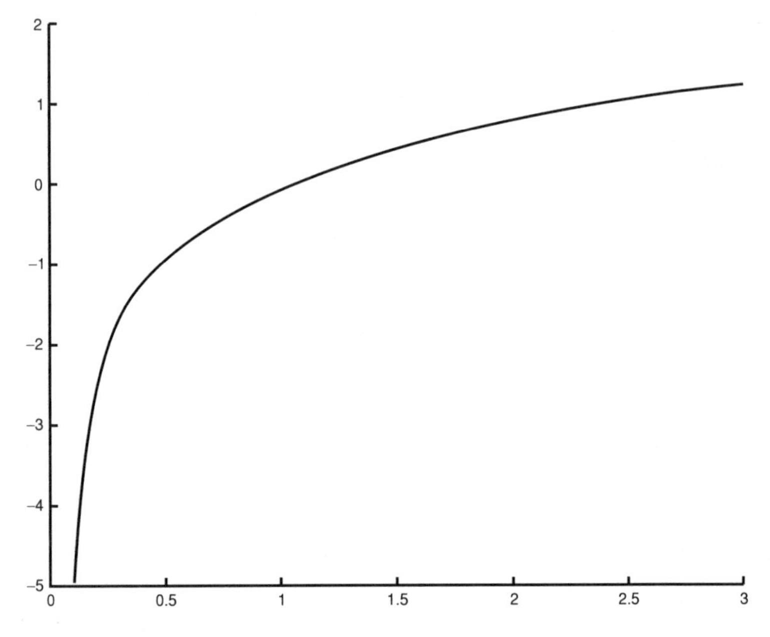

Figure 4. Solution of $y' + \frac{y}{x} = 1,\ y(1) = 0$.

on the interval $x > 0$. Multiply the equation by the integrating factor $\exp(\int^x \frac{ds}{s}) = \exp(\ln|x|) = x$ (up to an irrelevant nonzero multiplicative constant) to get the exact equation

$$x\frac{dy}{dx} + y = x$$

or $\frac{d}{dx}(xy) = x$. Integrating, $xy = x^2/2 + C$, and since $y(1) = 0$, we determine C and the unique solution

$$y(x) = \frac{1}{2}\left(x - \frac{1}{x}\right) \qquad \text{for } x > 0,$$

which is shown in Figure 4. We cannot reasonably expect to extend the solution to negative values of x (even though the formula for the solution is defined there) since the coefficient $1/x$ in the differential equation becomes unbounded at $x = 0$.

We will demonstrate through Example 14 that certain very special nonlinear equations can also be solved using the solution technique given for the general linear equation (1.30). Likewise, the equation $(\cos x)z' = 2x(\sin z - 1)$ is nonlinear for z, but it is linear for $y = \sin z$.

Example 14 The equation

$$\frac{dy}{dt} = \frac{1+y^2}{t+(1+y^2)}$$

is nonlinear in y, but linear in t, since it can be rewritten as

$$\frac{dt}{dy} = \frac{t}{1+y^2} + 1.$$

Using the integrating factor

$$e^{-\int^y \frac{dr}{1+r^2}} = e^{\tan^{-1} y},$$

we get the complicated-looking solution

$$t = e^{\tan^{-1} y} \int^y e^{-\tan^{-1} z} dz.$$

(Check it!) Finding the explicit inverse function $y(t)$ remains a challenge. Graphs of solutions would, anyway, perhaps be more helpful than a formula for $y(t)$.

Example 15 The equation

$$\frac{dy}{dt} = a(t)y + b(t)y^n \tag{1.33}$$

for some parameter n, not necessarily a positive integer, is a *Bernoulli equation*. The (already-encountered) special case $n = 2$ is called a Riccati equation and arises in very many applications; examples with $n = 0$ and 1 are linear, so they are already within grasp.

If we divide equation (1.33) by y^n and introduce $v = y^{1-n}$, then $\frac{dv}{dt} = \left(\frac{1-n}{y^n}\right)\frac{dy}{dt}$ and we obtain the linear equation

$$\frac{dv}{dt} = (1-n)a(t)v + (1-n)b(t)$$

for v. Thus, we can explicitly find $v(t)$ using the integrating factor $\exp[-(1-n)\int^t a(s)ds]$, presuming a and b are smooth, and then

$$y(t) = v^{1/(1-n)}.$$

Recall the population model

$$\frac{dN}{dt} = r\left(1 - \frac{N}{K}\right)N$$

with positive parameters r, K, and $N(0)$. Dividing by N^2 shows that $M \equiv 1/N$ satisfies $\frac{dM}{dt} + rM = \frac{r}{K}$ or $\frac{d}{dt}(e^{rt}M) = \frac{r}{K}e^{rt}$. Upon integration, we get

$$\frac{K}{N} = KM = 1 + Ce^{-rt}$$

with $C = \frac{K}{N(0)} - 1$. Thus, as found earlier,

$$N(t) = \frac{KN(0)}{N(0) + (K - N(0))e^{-rt}}.$$

Likewise, since the nonhomogeneous Riccati equation

$$u' + x^3u - x^2u^2 = 1$$

has the solution $u = x$, introducing

$$y = u - x$$

gives the homogeneous Riccati equation

$$y' = x^3y + x^2y^2$$

for y. Since this provides a linear equation for $z = 1/y$, it enables us to find more general solutions u.

1.5 Equations of homogeneous type

Equations of the form

$$\frac{dy}{dx} = F(y/x) \tag{1.34}$$

are said to be of *homogeneous type*. (There is no obvious connection between the two uses of the term homogeneous that we have encountered. Both usages are historic, which is unfortunate.) These equations can be solved by introducing the quotient

$$v = y/x \tag{1.35}$$

as a new variable. Since $y = xv$, $\frac{dy}{dx} = x\frac{dv}{dx} + v$ implies the separable equation

$$x\frac{dv}{dx} = F(v) - v,$$

which has the implicit solution

$$\ln|x| = \int^x \frac{ds}{s} = \int^v \frac{dz}{F(z) - z},$$

presuming the latter integral is defined. Thus,

$$x = Ke^{\int^{v} \frac{dz}{F(z)-z}}$$

for an arbitrary constant K. In practice, one can often find a more pleasing explicit representation of $y(x) = xv(x)$. We realize that it may take some experience to recognize equations as being of homogeneous type. Note, for example, that Examples 10, 11, and 13, considered above, are homogeneous.

Example 16 Consider

$$\frac{dy}{dx} = \frac{y^2 + 2xy}{x^2} = \left(\frac{y}{x}\right)^2 + 2\left(\frac{y}{x}\right).$$

Setting $y = xv$, we transform to $x\frac{dv}{dx} + v = v^2 + 2v$. v then satisfies the separable equation $\frac{dv}{v(v+1)} = \frac{dx}{x}$. By partial fractions, $\frac{1}{v(v+1)} = \frac{1}{v} - \frac{1}{v+1}$, so

$$\frac{dv}{v} - \frac{dv}{v+1} = \frac{dx}{x},$$

provided (perhaps) $v \neq 0$ or -1 and $x \neq 0$, and thus

$$\ln|v| - \ln|v+1| = \ln|x| + k$$

for a constant k. Exponentiating,

$$\frac{v}{x(v+1)} = \pm e^k \equiv c, \text{ so } \frac{y}{y+x} = cx$$

for some constant c. It follows that

$$y(x) = \frac{cx^2}{1-cx}, \quad \text{provided } x \neq \frac{1}{c}.$$

The validity of this solution can, of course, be directly checked, allowing us to stop worrying about the earlier transformations. Note that c would be directly determined by the initial condition $y(x_0) = y_0$, provided $x_0 \neq 0$ and $y_0 + x_0 \neq 0$.

1.6 Special second-order equations

Equations of the form

$$\frac{d^2y}{dt^2} = h\left(y, \frac{dy}{dt}\right) \tag{1.36}$$

are called *autonomous* when h does *not* depend directly on the independent variable t. We can interpret (1.36) as a differential equation whose solutions provide the position $y(t)$ of a body that moves according to Newton's law for a special kind of forcing h.

One, perhaps unexpected, way to solve (1.36) is to consider the velocity

$$V(y) \equiv \frac{dy}{dt}$$

as a function of y, rather than t, noting that this determines the acceleration as

$$\frac{d^2y}{dt^2} = \frac{dV}{dy}\frac{dy}{dt} = V\frac{dV}{dy}.$$

Thus, the task of solving the second-order equation (1.36) is reduced to solving the first-order equation

$$V\frac{dV}{dy} = h(y, V). \tag{1.37}$$

If we can solve this transformed equation for $V(y)$, the remaining separable first-order differentiate equation $V(y) = \frac{dy}{dt}$ allows us to implicitly obtain $y(t)$ from

$$t = \int^y \frac{dz}{V(z)}.$$

We will reconsider equation (1.36) with h independent of $\frac{dy}{dt}$ in Chapter 5 where we will provide a graphical solution technique based on the principle of conservation of energy.

Example 17 A simple application of the preceding approach is to study the linear frictionless spring whose deflection y satisfies the initial value problem

$$m\frac{d^2y}{dt^2} + ky = 0$$

for a positive mass m and a positive spring constant k, with prescribed initial values $y(0) = y_0$ and $\frac{dy}{dt}(0) = v_0$ for the position and velocity. Introducing $V(y) = \frac{dy}{dt}$ gives the transformed (separable and exact) equation $mV\frac{dV}{dy} = -ky$ or

$$mV\,dV + ky\,dy = 0.$$

Integration now conserves the sum of the kinetic and potential energies as

$$\frac{m}{2}V^2 + \frac{k}{2}y^2 = C \equiv \frac{m}{2}v_0^2 + \frac{k}{2}y_0^2$$

and, thereby, the velocity

$$\frac{dy}{dt} = V(y) = \pm\sqrt{\frac{k}{m}}\,\sqrt{\alpha^2 - y^2},$$

where $\alpha^2 \equiv y_0^2 + \frac{m}{k}v_0^2$ is specified. (The sign of V is uniquely determined by the prescribed initial velocity $v_0 = V(y_0)$). Thus, there remains only the separated equation

$$\frac{dy}{\sqrt{\alpha^2 - y^2}} = \pm\sqrt{\frac{k}{m}}dt,$$

for $y(t)$. Its general solution is

$$\sin^{-1}\frac{y}{\alpha} = \pm\sqrt{\frac{k}{m}}t + L$$

for some (yet unspecified) constant L, so the deflection must be given by

$$y(t) = \alpha \sin\left(\pm\sqrt{\frac{k}{m}}t + L\right) = (\pm\alpha\cos L)\sin\left(\sqrt{\frac{k}{m}}t\right) + (\alpha\sin L)\cos\left(\sqrt{\frac{k}{m}}t\right)$$

or as

$$y(t) = c_1 \sin\left(\sqrt{\frac{k}{m}}t\right) + c_2\cos\left(\sqrt{\frac{k}{m}}t\right)$$

for arbitrary constants c_1 and c_2. Applying the initial conditions, we determine c_1 and c_2 (and thereby L) uniquely and obtain the familiar equation

$$y(t) = y_0\cos\left(\sqrt{\frac{k}{m}}t\right) + v_0\sqrt{\frac{m}{k}}\sin\left(\sqrt{\frac{k}{m}}t\right)$$

completely describing *simple harmonic motion*.

Example 18 The nonlinear equation

$$\frac{d^2y}{dx^2} = \sqrt{1 + \left(\frac{dy}{dx}\right)^2}$$

describes the position y of a suspension cable, which either supports a bridge or hangs under its own weight. The equation is of the form (1.36), but it is also representative of those special second-order equations

$$\frac{d^2y}{dx^2} = f\left(x, \frac{dy}{dx}\right) \tag{1.38}$$

where the right-hand side f is independent of y and the first-order equation $\frac{du}{dx} = f(x, u)$ for $u = \frac{dy}{dx}$ can be directly solved to provide $y(x) = \int^x u(s)ds$. Separating variables in our example gives $dx = \frac{du}{\sqrt{1+u^2}}$, which implies that $x - x_0 = \ln(u + \sqrt{1+u^2})$ for some constant x_0. Thus, $u + \sqrt{1+u^2} = e^{x-x_0}$, and therefore $u - \sqrt{1+u^2} = e^{-(x-x_0)}$, and so

$$u = \frac{dy}{dx} = \frac{1}{2}\left(e^{x-x_0} + e^{-(x-x_0)}\right) \equiv \sinh(x - x_0).$$

Integration yields the convenient equation

$$y(x) = \cosh(x - x_0) + c$$

for the *catenary*, where x_0 is the location of the zero of y' (i.e., the minimum of y) and the constant of integration c is 0 if that minimum value is 1.

1.7 Epilog

The preceding techniques are very useful, but tend to give the incorrect impression that clever men and women need only use a few tricks to solve most scalar first-order differential equations. The extent of our known ability is, perhaps, exhibited in Kamke (1943). The more accessible treatment of Zwillinger (1989), which is in English, is also worth consulting. Explicit solutions are rare. This is why approximate methods of solution are, in practice, essential. Nonetheless, the author hopes that readers have enjoyed the challenge of trying to explicitly solve first-order equations and that they will do a good job on the exercises that follow. As in evaluating integrals or doing crossword puzzles, the needed skills are largely acquired from experience. There's no alternative!

1.8 Exercises

1 Solve the following initial value problems. Plot each solution and specify its interval of existence.

(a) $\frac{d^2x}{dt^2} = t^2, \; x(0) = 1, \; \dot{x}(0) = 2$

(b) $x^2\frac{dy}{dx} + y^2 = 0, \; y(1) = 3$

(c) $\dfrac{dy}{dx} = \dfrac{y^3 + 2y}{x^2 + 3x}$, $y(1) = 1$

(d) $y' + \dfrac{2}{x} y = \dfrac{\cos x}{x^2}$, $y(\pi) = 0$

(e) $y' = xye^{x^2}$, $y(0) = 1$. Explain why this differential equation guarantees that its solution is symmetric about $x = 0$.

2 Solve the following equations:

(a) $\dfrac{dx}{dt} + (\cos t)e^x = 0$

(b) $\dfrac{dz}{dr} = r^2(1 + z^2)$

(c) $y' = (4x - y)^2$ (Hint: Why not solve for $u = 4x - y$?)

(d) $(x + 1)y' = y + x$

(e) $2x^2 \dfrac{dy}{dx} = x^2 + y^2$

(f) $y' + \dfrac{a}{x} y = f(x)$

(g) $xy' = y + \sqrt{xy}$

(h) $y' + y + \frac{1}{y} = 0$

(i) $(x + 2) \sin y dx + x \cos y dy = 0$ (Hint: Try the integrating factor xe^x.)

(j) $y'' + y' = 3x^2$ (Hint: Introduce $v = y'$.)

(k) $xy' - y = \sqrt{x^2 + y^2}$

3 (a) Find a second-degree polynomial solution $y_p(x)$ of the linear equation

$$y' + 2y = x^2 + 4x + 7.$$

(b) Find the most general solution (depending, say, on an arbitrary initial value $y(0)$). (Hint: Find a differential equation for $z = y - y_p(x)$.)

4 Suppose a given function $f(x)$ satisfies the nonhomogeneous Riccati equation

$$\frac{dy}{dx} = a(x)y^2 + b(x)y + c(x).$$

Find its general solution by introducing $z = y - f$.

5 (a) Solve $xu' = u$.

(b) Using the solution $u(x)$ of (a), solve $z' - z = u$.

(c) Writing the linear variable coefficient second-order equation

$$xy'' - (1+x)y' + y = 0$$

as the *operator equation*

$$[xD^2 - (1+x)D + 1]y = 0$$

with $D = d/dx$, explain how the *factorization*

$$xD^2 - (1+x)D + 1 = (xD - 1)(D - 1)$$

and the result of (b) yields the solution $y(x)$.

6 Suppose the position of a body of mass m that is shot up from the surface of the Earth satisfies the inverse square law

$$m\ddot{x} = -mg\frac{R^2}{x^2},$$

where x is the distance from the Earth's center, R is the Earth's radius, and the initial conditions are

$$x(0) = R > 0, \ \dot{x}(0) = v_0 > 0.$$

If $v_0^2 \geq 2gR$, show that the body will never return to earth. The solution, thereby, defines the *escape velocity* for the body. (Hint: Determine the expression for $v(x)$ and its sign.) Note that this concept may explain why lighter elements escape the Earth's atmosphere.

7 Solve $(y')^2 + (x+y)y' + xy = 0$. (Hint: Solve for y'.)

8 The equation

$$y' = g' + \lambda(y - g),$$

for a given coefficient function $g(t)$ and a large negative parameter λ, is often used as a model of *stiff differential equations.*

(a) Solve the initial value problem on $t \geq 0$.

(b) Describe the limiting solution for $t > 0$ as $\lambda \to -\infty$.

9 Consider the equation

$$\frac{dy}{dx} = \frac{x + 2y + 4}{2x + y + 5}.$$

(a) Obtain a constant solution $(x, y) = (a, b)$.

(b) Introduce $X = x - a$, $Y = y - b$ and solve the resulting equation for $Y(X)$.

(c) Find the implicit solution when $y(0) = 0$.

10 (a) Solve

$$\frac{dy}{dx} = -e^{y+x+1}, \ y(0) = -1.$$

(b) Explain why the existence-uniqueness theorem guarantees that the initial value problem for $z = y + x + 1$ has only the trivial solution $z(x) \equiv 0$.

11 Solve

$$\frac{dy}{dx} + \left(\frac{x}{1+x^2}\right) y = (1+x^2)^{1/2}, \ y(0) = 5.$$

12 Determine an integral that satisfies

$$y' + \left(\frac{x}{1+x^2} + \frac{x^3}{2+x^4}\right) y = 1.$$

13 Show that $y = x$ cannot satisfy

$$y' + p(x)y^2 = 0$$

for any coefficient $p(x)$ that is continuous at $x = 0$.

14 Find the general solution to

$$y' = -y^2 + (1 - 2x)y + 2x,$$

given the special solution $y = 1$. (Hint: See exercise 4.)

15 Verify that the initial value problem

$$y' = -x + \sqrt{x^2 + 2y}, \ y(x_0) = -x_0^2/2$$

has the solution $y = -\frac{1}{2}x^2$ for all x and the additional solution $y = x_0\left(\frac{1}{2}x_0 - x\right)$ for $x \geq x_0$. Explain why the uniqueness theorem is not violated by examining $\frac{\partial}{\partial y}[-x + \sqrt{x^2 + 2y}]$ at the initial point.

16 Show that

(a) the expression $xy^2 + e^{-y} + 1$ seems to define an implicit solution $y(x)$ of the differential equation

$$(xy^2 + 2xy - 1)y' + y^2 = 0.$$

(b) no real solutions are so defined for $x \leq 0$, a positive solution exists for $x > 0$, and two negative solutions also occur for $x \gtrsim 2.07$ (use graphical methods to confirm this).

17 Determine the solution of the linear initial value problem

$$y' = (\sec^2 t)y, \quad y(2\pi) = 1.$$

Note that it is only defined for $\frac{3\pi}{2} \leq t < \frac{5\pi}{2}$, reflecting the singularities of the secant.

18 Solve

$$x(y-1)\frac{dy}{dx} = \ln x, \ y(1) = 1$$

and determine the interval of existence.

19 Show that all solutions x of

$$\frac{dx}{dt} = \frac{x}{t} + \sqrt{x}$$

tend to t^2 as $t \to \infty$.

20 Solve the initial value problem for

$$(xy)' = xy' + y = x \sin x, \quad \text{with } y(1) \text{ given.}$$

Show that solutions become unbounded at the singular point $x = 0$ unless

$$\lim_{x \to 0}(xy) = 0.$$

21 Show why the autonomous equation

$$yy'' = (y')^2$$

has solutions $y = ce^{kx}$ for arbitrary constants c and k.

22 Solve

$$yy'' = (y')^2 - y', \ y(0) = 1, \ y'(0) = 2.$$

23 Integrate

$$y'' + y + y^3 = 0$$

once and interpret the result in terms of conservation of energy.

(a) Explain why any solution is bounded.

(b) Look up *elliptic integrals* (in a mathematical handbook) and express the solutions in terms of them.

24 The nonlinear equation

$$\ddot{\theta} + p(\dot{\theta})^2 + q \sin\theta = 0,$$

for positive constants p and q, is often used to model the motion of a pendulum. Show that it provides a linear equation for ω^2, where $\omega(\theta) \equiv \dot{\theta}$. Use ω to find θ implicitly.

25 Consider *Clairaut's equation*

$$xy' + f(y') = y.$$

Differentiation implies that $y''(x + f'(y')) = 0$, suggesting the family of straight line solutions

$$y = mx + f(m)$$

for an arbitrary slope m, as well as the singular solution (or "envelope")

$$\begin{cases} x = -f'(m) \\ y = -mf'(m) + f(m) \end{cases}$$

parameterized by m.

Use this approach to find the singular solution of

$$2xy' - (y')^2 = y.$$

26 Consider the second-order scalar equation

$$z'' + g(z, z', x) = 0$$

with prescribed initial values $z(x_0)$ and $z'(x_0)$. By introducing the vector

$$y = \begin{pmatrix} z \\ z' \end{pmatrix}$$

and assuming the existence-uniqueness theorem for the vector initial value problem

$$y' = f(x, y), \;\; y(x_0) = y_0,$$

show that the given problem has a unique solution locally when g, g_z, and $g_{z'}$ are all continuous near $(z(x_0), z'(x_0), x_0)$.

27 Show that the linear initial value problem

$$xy' = 2(y - 1), \quad y(0) = 1$$

has the smooth solutions

$$y(x) = \begin{cases} ax^2 + 1, & x < 0 \\ bx^2 + 1, & x > 0 \end{cases}$$

for arbitrary constants a and b, but no solution for $y(0) \neq 1$. Does this violate the uniqueness theorem?

28 Show that the separable equation

$$yy' = (y + 1)^2$$

has the solutions $y = -1$ and

$$\frac{1}{y+1} + \ln|y + 1| = x + c.$$

29 Solve

$$\frac{d^2y}{dx^2} = e^{1-\frac{dy}{dx}}$$

30 Suppose a fluid of height h flows out of a basin with cross-sectional area $A(h)$ through a small hole of area a at $h = 0$. Pressure balance at the hole requires that

$$\tfrac{1}{2}\rho v^2 = \rho g h,$$

where ρ is the density of the fluid, v is the velocity, and g is the acceleration of gravity, so $v = \sqrt{2gh}$ for $h \geq 0$. Balancing the volume change, then provides the separable differential equation

$$A(h)\frac{dh}{dt} = -a\sqrt{2gh}.$$

Find $h(t)$ for typical areas $A(h)$.

31 Solve the two-point problem

$$\begin{cases} y'' = y, & 0 \leq x \leq 1 \\ y(0) + 3y(1) = 1, & y'(1) = 0 \end{cases}$$

in terms of hyperbolic functions.

32 (a) Find a solution $y_p(x)$ of

$$y' + y = \sin x$$

of the form $y_p(x) = \alpha \sin x + \beta \cos x$.

(b) Find a solution of the initial value problem $y' + y = \sin x$, $y(0)$ prescribed of the form $y(x) = y_p(x) + \gamma e^{-x}$ for $x \geq 0$.

(c) Explain why all solutions of the initial value problem have a long-time limit which is independent of the prescribed initial value $y(0)$.

33 The current flowing in an electric circuit (with negligible capacitance) satisfies

$$L\frac{di}{dt} + Ri = v_0 \sin \omega t$$

for positive constants L, R, v_0, and ω. Find the current $i(t)$ for $t > 0$, given $i(0) = 0$.

34 Show that $f(x) = \frac{f(0)}{(1+cx)^2}$ is the general solution of

$$\frac{1}{x}\int_0^x f(t)dt = \sqrt{f(0)f(x)}$$

by finding $f'(x)$.

35 Use solutions of the complex Riccati equation

$$\dot{z} = z^2 + iz$$

to solve the real nonlinear system

$$\dot{x} = -y + x^2 - y^2$$
$$\dot{y} = x + 2xy$$

for $z = x + iy$. Verify your answer.

36 (a) Show that the linear equation

$$x^2y' - y = -x$$

has the solution

$$y(x) = e^{-\frac{1}{x}}\int_{-\infty}^{\frac{1}{x}} \frac{e^s}{s}ds \quad \text{for } x < 0.$$

(b) Determine polynomial approximations to this solution, for x small, using integration by parts.

(c) Equate coefficients formally in the differential equation to find a power series "solution" of the form

$$y(x) = \sum_{n=1}^{\infty} a_n x^n.$$

37 Let $\theta(t)$ be defined implicitly through the transcendental equation

$$t = 2K(k, \theta/2)$$

where

$$K(k,\theta) \equiv \int_0^\theta \frac{dt}{\sqrt{1 - k^2 \sin^2 t}}$$

defines an *elliptic integral* of the first kind. Show that θ satisfies

$$(\dot{\theta})^2 = 1 - \tfrac{1}{2}k^2(1 - \cos\theta).$$

38 Show that the linear integral equation

$$f(t) = g(t) + \int_0^t K(s) f(s) ds$$

has the unique solution

$$f(t) = g(t) + \int_0^t K(s) e^{\int_s^t K(r)dr} g(s) ds$$

under the assumptions that g is continuous and K is integrable. Show that the solution could be obtained using successive approximations and that the solution is the unique solution of the initial value problem obtained by differentiating the integral equation, when that is allowed.

39 Show that

$$\frac{x dy - y dx}{x^2 - y^2} = 0$$

is exact with solutions $\frac{x+y}{x-y} = c$.

40 Find two real nonzero solutions $y(x)$ of $y' + xy = x/y^3$.

41 Find a two-parameter family of solutions of the second-order equation

$$y'' + 6y(y')^3 = 0.$$

42 Because the solution of

$$\frac{dy}{dt} = y^2 + t^2, \quad y(0) = 1$$

grows faster than that of

$$\frac{dz}{dt} = z^2, \quad z(0) = 1,$$

explain why y blows up for some positive t.

43 Use the linear system

$$y' = \begin{pmatrix} 0 & 1 \\ 1 & 0 \end{pmatrix} y$$

and the existence-uniqueness theorem for vector systems to show that $\cos ix = \cosh x$.

44 Use conservation of energy to show that any solution to

$$y'' + y + y^3 = 0$$

must remain bounded (so it continues to exist for all x). (Note that y might represent the displacement of a "hard" spring.)

45 Consider the *delay-differential equation*

$$y'(x) = f(x, y(x - \tau))$$

for a smooth function f and a constant *delay* $\tau > 0$, and suppose a smooth initial function $y(x)$ is prescribed on the interval $-\tau \leq x \leq 0$.

(a) Show that the solution satisfies the integral equation

$$y(x) = y(0) + \int_0^x f(s, y(s - \tau))ds$$

for y on $x \geq 0$.

(b) Use the integral equation to define the solution $y(x)$ on the successive intervals $j\tau \leq x \leq (j + 1)\tau$, $j = 0, 1, 2, \ldots$.

46 A tank initially contains 98 gallons of pure water. Starting at time $t = 0$, a supply containing one pound of pollutant per gallon flows into the tank at a rate of two gallons per minute. After the first minute, the stirred mixture continues to flow out of the tank at a rate of two gallons per minute.

(a) How much pollutant and how many gallons of water are in the tank after the first minute?

(b) How much pollutant is in the tank after one hour?

(c) How much pollutant remains as $t \to \infty$?
(Idea: If P is the amount of pollutant, $\frac{dP}{dt}$ will be the difference between its inflow and outflow.)

47 (a) Solve the scalar initial value problem

$$\frac{dx}{dt} = a(t)x, \; x(0) = 1.$$

(b) Show that the solution of

$$\frac{dx}{dt} = (\sin t)x, \; x(0) = 1$$

has period 2π.

(c) Suppose $a(t)$ has period p (i.e., $a(t+p) = a(t)$ for all t and some $p > 0$) and let

$$z(t) = e^{-\alpha t} x(t),$$

where $x(t)$ is the solution of (a). Show that $z(t)$ has period p if α is the average $\frac{1}{p}\int_0^p a(s)ds$ of $a(t)$ over one period.

(d) Also show that the solution of (c) satisfies

$$x(t+p) = cx(t) \text{ for } c = e^{\alpha p}.$$

What condition on α guarantees that $x(t) \to 0$ as $t \to \infty$?

(e) Show that the solution of

$$\frac{dx}{dt} = (\sin t)(x+1), \;\; x(0) = 1$$

has period 2π,whereas that of

$$\frac{dx}{dt} = (\sin t)x + 1, \;\; x(0) = 1$$

does not.

(f) Show that the solution of

$$\frac{dx}{dt} = a(t)x + b(t), \;\; x(0) = 1$$

has period p when $\int_0^p e^{-\gamma(s)} b(s)ds = 0$ for $\gamma(t) = \int_0^t a(r)dr$, presuming the homogeneous equation has a solution of period p.

48 Write an essay on the topic: All solvable first-order scalar ordinary differential equations can be solved as exact equations.

Solutions to exercises 1–6

1 (a) $\dfrac{d^2x}{dt^2} = t^2$, $x(0) = 1$, $\dot{x}(0) = 2$. Successive integrations imply that $\dfrac{dx}{dt} = \dfrac{t^3}{3} + 2$ and $x(t) = \dfrac{t^4}{12} + 2t + 1$.

(b) $x^2\dfrac{dy}{dx} + y^2 = 0$, $y(1) = 3$. The equation is separable, yielding $\dfrac{dy}{y^2} + \dfrac{dx}{x^2} = -d\left(\dfrac{1}{y} + \dfrac{1}{x}\right) = 0$, so $\dfrac{1}{y} + \dfrac{1}{x} = \dfrac{4}{3}$ due to the initial condition and $y(x) = \dfrac{1}{\frac{4}{3} - \frac{1}{x}}$ for $x > 3/4$.

(c) $\dfrac{dy}{dx} = \dfrac{y^3+2y}{x^2+3x}$, $y(1)=1$. Rewriting the separable equation as $\dfrac{1}{2}\dfrac{dy}{y} - \dfrac{1}{4}\left(\dfrac{2ydy}{y^2+2}\right) = \dfrac{1}{3}\left(\dfrac{dx}{x} - \dfrac{dx}{x+3}\right)$, integration yields $\ln(y^2) - \ln(y^2+2) = \frac{4}{3}(\ln|x| - \ln|x+3|) + \ln c$. Thus, $\dfrac{y^2}{y^2+2} = c\left(\dfrac{x}{x+3}\right)^{4/3}$, where the initial condition provides $c = \frac{1}{3}4^{1/3}$. The solution $y = \sqrt{\frac{2}{3\left(\frac{x+3}{4x}\right)^{3/4}-1}}$ is defined for $\dfrac{-3^{7/4}}{4+3^{3/4}} < x < \dfrac{3^{7/4}}{4-3^{3/4}}$.

(d) $y' + \frac{2}{x}y = \frac{\cos x}{x^2}$, $y(\pi)=0$. The linear equation has the integrating factor $\exp(2\int^x \frac{ds}{s}) = x^2$, so it can be rewritten as $\frac{d}{dx}(x^2y) = \cos x$. Integration yields $x^2y = \sin x$ since $y(\pi)=0$. Thus, $y(x) = \frac{\sin x}{x^2}$ for $x>0$.

(e) $y' = 2xye^{x^2}$, $y(0)=1$. Integrating the separated linear equation $y'/y = 2xe^{x^2}$ yields $\frac{1}{2}\ln y^2 = e^{x^2}-1$. Thus, $y(x) = e^{(e^{x^2}-1)}$. If we change x to $s=-x$, $\dfrac{dy}{dx} = -\dfrac{dy}{ds}$ implies that $y(s)=y(x)$, i.e., $y(x)=y(-x)$ is symmetric.

2 (a) $\dfrac{dx}{dt} + (\cos t)e^x = 0$ implies that $e^{-x}dx + \cos t\,dt = 0$, so integrating yields $-e^{-x} = -\sin t + c$ and $x = -\ln|c-\sin t|$ as long as it is defined.

(b) $\dfrac{dz}{dr} = r^2(1+z^2)$ is separable, so $\dfrac{dz}{1+z^2} = r^2dr$ integrates to $\tan^{-1}z = \frac{r^3}{3}+c$, yielding $z(r) = \tan\left(\frac{r^3}{3}+c\right)$.

(c) $y' = (4x-y)^2$. If we introduce $u = 4x-y$, we get $u' = 4 - y' = 4-u^2$, so $\frac{du}{4-u^2} = dx$ implies that $\frac{1}{2}\tanh^{-1}\frac{u}{2} = x + \frac{c}{2}$. Thus, $u = 2\tanh(2x+c)$ and $y = 4x - 2\tanh(2x+c)$.

(d) $(x+1)y' - y = x$ is linear, so we seek an integrating factor. Multiplying by $(x+1)^{-2}$, we get

$$\frac{d}{dx}\left(\frac{y}{x+1}\right) = \frac{y'}{x+1} - \frac{y}{(x+1)^2} = \frac{x}{(x+1)^2} = \frac{1}{x+1} - \frac{1}{(x+1)^2}.$$

Integrating now yields $\frac{y}{x+1} = \ln(x+1) + \frac{1}{x+1} + c$, so we get the general solution $y(x) = 1 + c(x+1) + (x+1)\ln(x+1)$.

(e) $2x^2\frac{dy}{dx} = x^2+y^2$ is of homogeneous form, so setting $y = vx$ implies $2v'x + 2v = 1+v^2$ and the separated equation $\frac{dv}{(v-1)^2} = \frac{dx}{2x}$. Integrating gives $-\frac{1}{v-1} = \ln\sqrt{|x|} + \ln k$, so $y = x - \frac{x}{\ln(k\sqrt{|x|})}$.

(f) $y' + \frac{a}{x}y = f(x)$ is linear with the integrating factor x^a, so $\frac{d}{dx}(x^ay) = x^af(x)$. Integration yields $y = x^{-a}(c + \int^x s^af(s)ds)$.

(g) $xy' = y + \sqrt{xy}$ is of homogeneous form (as well as Bernoulli). Setting $y = xv$ implies that $y' = xv' + v = v + \sqrt{v}$ or $\frac{dv}{\sqrt{v}} = \frac{dx}{x}$. Integrating, $2\sqrt{v} = \ln|x| + c$ or $y = \frac{x}{4}(\ln|x| + c)^2 = \frac{x}{4}\ln(kx)$ for some constant k. Note that $y = 0$ is also a solution, though it does not correspond to any special k.

(h) $y' + y + 1/y = 0$ is separable (and Bernoulli). Setting $\frac{2ydy}{1+y^2} = -2dx$, integration yields $\ln(1 + y^2) = -2x + \ln c^2$ or $y = \pm\sqrt{-1 + c^2e^{-2x}}$ as long as it is defined.

(i) $(x + 2)\sin y dx + x\cos y dy = 0$ becomes exact after multiplication by the integrating factor xe^x. (It is also separable.) Thus, $\frac{d}{dx}(x^2e^x \sin y) = 0$ implies that $\sin y = \frac{c}{x^2e^x}$ or $y = \sin^{-1}(cx^{-2}e^{-x})$.

(j) $y'' + y' = 3x^2$ implies the linear first-order equation $y' + y = x^3 + c$ upon integration. Thus, $(ye^x)' = e^x(x^3 + c)$ implies that $ye^x = e^x(x^3 - 3x^2 + 6x + k_1) + k_2$ for any constants k_1 and k_2 or $y = k_1 + k_2e^{-x} + x^3 - 3x^2 + 6x$.

(k) $xy' - y = \sqrt{x^2 + y^2}$ is of homogeneous form, so $y = vx$ implies that $\frac{dv}{\sqrt{1+v^2}} = \frac{dx}{x}$ and $\sinh^{-1} v = \ln|x| + c$ yields $y = x\sinh(\ln|x| + c) = \frac{x}{2}(ke^{\ell n|x|} - \frac{1}{k}e^{-\ln|x|}) = \frac{\operatorname{sgn} x}{2k}(k^2x^2 - 1)$ or $y = \frac{1}{2c}(c^2x^2 - 1)$ for $x > 0$ or $x < 0$ and an arbitrary constant c.

3 (a) If we seek a quadratic solution $y_p = Ax^2 + Bx + C$ of $y' + 2y = x^2 + 4x + 7$, substitution requires that $(2Ax + B) + 2(Ax^2 + Bx + C) = x^2 + 4x + 7$, so $2A = 1$, $2A + 2B = 4$, and $B + 2C = 7$. Thus, $y_p = \frac{1}{2}x^2 + \frac{3}{2}x + \frac{11}{4}$.

(b) Because $z = y - y_p(x)$ satisfies $z' + 2z = e^{-2z}\frac{d}{dx}(ze^{2x}) = 0$, we will have $z = e^{-2x}c$ for some constant c, so $y = e^{-2x}c + \frac{1}{2}x^2 + \frac{3}{2}x + \frac{11}{4}$ for $c = y(0) - \frac{11}{4}$.

4 If $y' = a(x)y^2 + b(x)y + c(x)$ and $f' = a(x)f^2 + b(x)f + c(x)$, let $z = y - f$ so $y = z + f$ must satisfy the given Riccati equation, which implies that z will satisfy $z' = az^2 + (2af + b)z$. Dividing by z^2, we have $\frac{d}{dx}\left(\frac{1}{z}\right) = -a + (2af + b)/z$. Integration of this linear equation for $1/z$ yields $\frac{1}{z} = -\int^x \exp[-\int_t^x [2a(s)f(s) + b(s)]ds]a(t)dt$ and $y = z(x) + f(x)$.

5 (a) If $xu' = u$, $\frac{du}{u} = \frac{dx}{x}$ implies that $u = kx$ for some constant k.

(b) If $z' - z = u = kx$, another integration implies that $(ze^{-x})' = kxe^{-x}$ or $z(x) = ce^x - k(1 + x)$ for arbitrary constants c and k.

(c) We can write the linear second-order equation $xy'' - (1 + x)y' + y = 0$ as the operator equation $[xD^2 - (1 + x)D + 1]y = 0$ using the derivative operator $D = \frac{d}{dx}$. Moreover, we can factor the operator as $xD^2 - (1 + x)D + 1 = (xD - 1)(D - 1)$. By artificially introducing $(D - 1)y = y' - y = u$, the given equation becomes $(xD - 1)u$

$= xu' - u = 0$. We can then directly check that the z found in (b) satisfies the given equation for y.

6 The equation $m\ddot{x} = -\frac{mgR^2}{x^2}$ is second-order, but autonomous. Considering $v = \dot{x}$ as a function of x, rather than t, we get $v\,dv = -\frac{gR^2}{x^2}dx$, so integration yields $\frac{v^2}{2} = \frac{gR^2}{x} + \frac{v_0^2}{2} - gR$ because $x(0) = R$ and $v(R) = v_0$. Since $v_0 > 0$, $v = \left[\frac{2gR^2}{x} + (v_0^2 - 2gR)\right]^{1/2}$. If $v_0^2 \geq 2gR$, $\frac{2gR^2}{x} > 0$ implies that the velocity v is always positive, so the distance $x - R$ from the earth increases indefinitely with $t = \int_R^x \left[\frac{2gR^2}{s} + (v_0^2 - 2gR)\right]^{-1/2} ds$. For smaller values of v_0, we would have to change the sign of v after it becomes zero to return the body to Earth.

2

Linear second-order equations

2.1 Homogeneous equations

At first, let us be primarily concerned with linear homogeneous scalar equations

$$y'' + a(x)y' + b(x)y = 0 \tag{2.1}$$

in standard form with continuous coefficients $a(x)$ and $b(x)$. The existence-uniqueness theorem guarantees that such equations have a unique solution when initial values $y(x_0)$ and $y'(x_0)$ are prescribed at some point x_0, while basic linear algebra can be used to describe the affine structure of the solution space for (2.1).

If $y_1(x)$ and $y_2(x)$ are two functions that both satisfy (2.1), substitution into the differential equation implies that any *linear combination*

$$y(x) = y_1(x)c_1 + y_2(x)c_2 \tag{2.2}$$

with arbitrary constants c_1 and c_2 will also satisfy (2.1). Indeed, we will assume that, though it is certainly not obvious, *all* solutions $y(x)$ of (2.1) have the form (2.2), provided $y_1(x)$ and $y_2(x)$ are any pair of *linearly independent* solutions, i.e., when the combination

$$y_1(x)c_1 + y_2(x)c_2 = 0 \tag{2.3}$$

is only true for the trivial set of constants $c_1 = c_2 = 0$.

Otherwise, if $y_1(x)c_1 + y_2(x)c_2 = 0$ and c_1, say, is nonzero, we will have $y_1(x)$ as the constant multiple $-y_2(x)\frac{c_2}{c_1}$ of $y_2(x)$, and we will naturally call y_1 and y_2 *linearly dependent*. Readers unfamiliar with such linear algebra concepts might refer, for example, to Bronson (1970) (among many possibilities), since linear algebra and matrix manipulations underlie much of our upcoming work.

Given two scalar functions, it is certainly easy to check whether or not one is a constant multiple of the other (by examining their quotient). Suppose we know two linearly independent solutions $y_1(x)$ and $y_2(x)$ of (2.1). Then, the initial (or, more generally, any other pair of appropriate *auxiliary*) conditions which define the solution $y(x)$ must be used to determine the unknowns c_1 and c_2 in the anticipated linear combination (2.2), i.e., the relations

$$\begin{cases} y(x_0) = y_1(x_0)c_1 + y_2(x_0)c_2 \\ \text{and} \\ y'(x_0) = y_1'(x_0)c_1 + y_2'(x_0)c_2 \end{cases}$$

and *Cramer's rule* will uniquely determine the previously unspecified constants as the quotients

$$c_1 = \begin{vmatrix} y(x_0) & y_2(x_0) \\ y'(x_0) & y_2'(x_0) \end{vmatrix} \Bigg/ \begin{vmatrix} y_1(x_0) & y_2(x_0) \\ y_1'(x_0) & y_2'(x_0) \end{vmatrix}$$

and

$$c_2 = \begin{vmatrix} y_1(x_0) & y(x_0) \\ y_1'(x_0) & y'(x_0) \end{vmatrix} \Bigg/ \begin{vmatrix} y_1(x_0) & y_2(x_0) \\ y_1'(x_0) & y_2'(x_0) \end{vmatrix},$$

provided the common denominator is nonzero. Thus, linear independence of solutions $y_1(x)$ and $y_2(x)$ of (2.1) must imply that $y_1(x_0)y_2'(x_0) \neq y_2(x_0)y_1'(x_0)$ for any prescribed initial point x_0.

More generally, let us briefly consider the nth-order linear homogeneous equation

$$y^{(n)} + a_1(x)y^{(n-1)} + \ldots + a_{n-1}(x)y' + a_n(x)y = 0 \tag{2.4}$$

(in standard form), with continuous coefficients $a_i(x)$ on some interval about an initial point x_0. One could use the existence-uniqueness theory to show that all solutions of (2.4) are given by the linear combination

$$y(x) = y_1(x)c_1 + y_2(x)c_2 + \ldots + y_n(x)c_n \tag{2.5}$$

for arbitrary constants $c_1, c_2, \ldots, c_n$ and any set $y_1(x), y_2(x), \ldots, y_n(x)$ of n linearly independent solutions of (2.4) that we might be fortunate enough to know. Most critically, we shall (somewhat magically) check the linear dependence or independence of any given set of n solutions $\{y_i(x)\}$ by calculating their *Wronskian determinant*

$$W(x) \equiv W[y_1, y_2, \ldots, y_n](x) = \begin{vmatrix} y_1(x) & y_2(x) & \ldots & y_n(x) \\ y_1'(x) & y_2'(x) & \ldots & y_n'(x) \\ \vdots & & & \\ y_1^{(n-1)}(x) & y_2^{(n-1)}(x) & \ldots & y_n^{(n-1)}(x) \end{vmatrix}. \tag{2.6}$$

If

$$\begin{cases} W(x) \neq 0, & \text{the } y_i\text{s are linearly independent at } x \\ W(x) = 0, & \text{the } y_i\text{s are linearly dependent there.} \end{cases} \tag{2.7}$$

The existence-uniqueness theory can be further used to show that a set of n linearly independent solutions $\{y_i(x)\}$ to the nth-order homogeneous equation (2.4) always exists; thus, the representation (2.5) of solutions implies that the collection of all solutions of (2.4) spans an n-dimensional linear vector space with $y_1(x), \ldots, y_n(x)$ as a *basis* (thereby, linear differential equations provide a very important and concrete motivation for studying linear algebra). To explicitly find n linearly independent solutions is not always possible, unless the a_js are constants or we are somehow lucky. If our collection of n solutions of (2.4) has a zero Wronskian, we have no choice but to seek more (linearly independent!) solutions to attain the general solution (2.5) and, thereby, all solutions. As for $n = 2$, finding the n coefficients c_i in the linear combination (2.5) from a prescribed set of initial values $y(x_0), y'(x_0), \ldots, y^{(n-1)}(x_0)$, given at some point x_0, will be uniquely possible only if the resulting Wronskian determinant $W(x_0)$ is nonzero.

Example 1 The linear homogeneous second-order equation

$$2x^2y'' + 3xy' - y = 0$$

can be written in the standard form

$$y'' + \frac{3}{2x}y' - \frac{1}{2x^2}y = 0$$

with continuous coefficients, as long as $x \neq 0$. It is easy to check that $y_1(x) = x^{1/2}$ and $y_2(x) = 1/x$ are linearly independent solutions for $x \neq 0$. (We might have cleverly found them by seeking solutions of the form x^p for some powers p to be determined.) We can check that the Wronskian

$$W(x) \equiv W[y_1, y_2](x) = \begin{vmatrix} x^{1/2} & \frac{1}{x} \\ \frac{1}{2x^{1/2}} & -\frac{1}{x^2} \end{vmatrix} = -\frac{3}{2}x^{-3/2} \neq 0,$$

or note that the ratio y_1/y_2 is not constant, so the general solution for $x \neq 0$ must be given by

$$y(x) = x^{1/2}c_1 + c_2/x$$

for arbitrary constants c_1 and c_2. This can also be verified by noting that the unique solution of the initial value problem $2x^2y'' + 3xy' - y = 0$, with $y(x_0)$ and $y'(x_0)$ prescribed, is

$$y(x) = \frac{2}{3}\left(\frac{x}{x_0}\right)^{1/2}(y(x_0) + x_0 y'(x_0)) + \frac{x_0}{3x}(y(x_0) - 2x_0 y'(x_0))$$

for $x > 0$ if $x_0 > 0$ (and for $x < 0$ if $x_0 < 0$). (Check it!) We need only solve two linear equations for c_1 and c_2 to get this unique answer. Note, however, the actual breakdown of this solution at the singular point $x = 0$ unless $y(x_0) = 2x_0 y'(x_0)$.

We will prove the Wronskian test (2.6) for $n = 2$. If $W[y_1, y_2](x) = 0$, $\frac{y_2'}{y_2} = \frac{y_1'}{y_1}$, so integration implies that $\ln|y_2| = \ln|y_1| + \ln|k|$ for some constant k or

$$y_2 = y_1 k,$$

i.e., we have linear dependence of y_1 and y_2. If, on the other hand, $W[y_1, y_2](x) \neq 0$ and, for example, $y_2 = y_1 k$ for some constant k, we would have

$$W[y_1, y_2](x) = \begin{vmatrix} y_1 & y_1 k \\ y_1' & y_1' k \end{vmatrix} = 0,$$

which would be a contradiction. Thus, a nonzero Wronskian implies linear independence of the functions being tested. Now suppose y_1 and y_2 are both solutions of (2.1). Then, differentiating

$$W(x) \equiv y_1 y_2' - y_1' y_2$$

implies that $W'(x) = y_1' y_2' + y_1 y_2'' - y_1' y_2' - y_1'' y_2 = y_1[-a(x)y_2' - b(x)y_2] - [-a(x)y_1' - b(x)y_1]y_2 = -a(x)(y_1 y_2' - y_1' y_2) = -a(x)W(x)$. Integrating the linear differential equation $W' + aW = 0$ then provides *Abel's formula*

$$W(x) = e^{-\int_{x_0}^{x} a(s)ds} W(x_0)$$

for any initial point x_0. Therefore, if y_1 and y_2 are linearly independent solutions of (2.1) (with a nonzero Wronskian), they will remain linearly independent as long as the continuity of $a(x)$ is maintained and the differential equation remains valid. The results are similar, for linear dependent solutions (i.e., $W(x_0) = 0$ for some x_0 implies that $W(x) = 0$ for all other x).

If we have two linearly independent solutions $y_1(x)$ and $y_2(x)$ of (2.1), it is easy to check that the initial value problem for (2.1) has the unique solution

$$Y(x) = y_1(x)\frac{W[y, y_2](x_0)}{W[y_1, y_2](x_0)} + y_2(x)\frac{W[y_1, y](x_0)}{W[y_1, y_2](x_0)}. \tag{2.8}$$

This is simply a highbrowed way of writing the unique constants c_1 and c_2 in (2.2), found previously via Cramer's rule. The alternative representation

$$y(x) = [y_1(x)y_2'(x_0) - y_2(x)y_1'(x_0)]\frac{y(x_0)}{W(x_0)} - [y_1(x)y_2(x_0) - y_2(x)y_1(x_0)]\frac{y'(x_0)}{W(x_0)}$$

emphasizes the fundamental role played by the initial values $y(x_0)$ and $y'(x_0)$.

Now we will show that we can find y_2, and thereby the general solution of (2.1), from just knowing one nontrivial solution $y_1(x)$ to (2.1). Because we want a solution y_2 that is linearly independent of y_1, it is natural to look for it in the form

$$y_2(x) = y_1(x)v(x), \tag{2.9}$$

where the quotient $v(x)$ is a nonconstant function to be determined. Differentiating twice, we get $y_2' = y_1'v + y_1v'$ and $y_2'' = y_1''v + 2y_1'v' + y_1v''$, so substitution into (2.1) implies that

$$(y''_1v + 2y_1'v' + y_1v'') + a(x)(y_1'v + y_1v') + b(x)y_1v = 0.$$

Since y_1 satisfies (2.1), however, the v coefficients cancel, so $v'' = \frac{dv'}{dx}$ implies that v' must satisfy the first-order linear equation

$$\frac{dv'}{dx} + \left(a(x) + \frac{2y_1'}{y_1}\right)v' = 0,$$

which we know how to solve. Using the (hopefully nonzero) integrating factor

$$\exp\left[\int^x \left(a(s) + \frac{2y_1'(s)}{y_1(s)}\right)ds\right] = e^{\int^x a(s)ds}e^{\ln(y_1^2(x))} = y_1^2(x)e^{\int^x a(s)ds},$$

we find that

$$\frac{d}{dx}\left(y_1^2(x)e^{\int^x a(s)ds}v'(x)\right) = 0,$$

so

$$v'(x) = \frac{1}{y_1^2(x)}e^{-\int^x a(s)ds}C$$

for an arbitrary constant C, which can be simply set equal to one due to the integral's compensating unspecified lower limit of integration. Thus, we immediately obtain

$$y_2(x) = y_1(x)\int^x y_1^{-2}(t)e^{-\int^t a(s)ds}dt$$

[if we (somewhat recklessly) don't worry about the possibility of integrating through zeros of y_1]. Since we can directly check that $y_2(x)$ does satisfy (2.1) and that $y_2(x)$ is not a constant multiple of $y_1(x)$, any solution of (2.1) must have the form

$$y(x) = y_1(x) \left[c_1 + \left(\int^x y_1^{-2}(t) e^{-\int^t a(s)ds} dt \right) c_2 \right] \tag{2.10}$$

for some constants c_1 and c_2. A similar *reduction of order* technique applies if we know any integer number m, $0 < m < n$, of linearly independent solutions of the nth-order linear homogeneous equation (2.4). (We will describe the more general corresponding result for systems of first-order equations in Chapter 4.) More significant is that an analogous procedure will provide us a *variation of parameters* method for solving all linear nonhomogeneous second-order differential equations, if we simply know one nontrivial solution of the corresponding homogeneous equation (see Section 2.2).

Example 2 Suppose we (somehow) guess the nontrivial solution $y_1(x) = e^{-x}$ of

$$y'' + 2y' + y = 0.$$

(We might have tried the possibility $e^{\lambda x}$ for an unspecified constant λ and found that $\lambda = -1$ will satisfy the *characteristic polynomial* $\lambda^2 + 2\lambda + 1 = 0$.) Then, we can seek the general solution using the ansatz (or guess)

$$y = e^{-x} v(x)$$

for a unknown variable function $v(x)$. Differentiating twice, we find that the equation for y requires

$$e^{-x}(v'' - 2v' + v) + 2e^{-x}(v' - v) + e^{-x} v = e^{-x} v'' = 0,$$

so we take $v(x) = c_1 + xc_2$ linear to get the general solution

$$y(x) = e^{-x}(c_1 + xc_2).$$

Example 3 Suppose $y_1(x) = x^2$ and $y_2(x) = x^3$ both satisfy some linear homogeneous differential equation $y'' + a(x)y' + b(x)y = 0$ for some unknown continuous coefficients $a(x)$ and $b(x)$. Because y_1 and y_2 are linearly independent, any solution y must then be given by

$$y(x) = x^2 c_1 + x^3 c_2$$

for some constants c_1 and c_2. But $y' = 2xc_1 + 3x^2c_2$ and $y'' = 2c_1 + 6xc_3$, so the *augmented Wronskian determinant*

$$W[y, x^2, x^3](x) \equiv \begin{vmatrix} y & x^2 & x^3 \\ y' & 2x & 3x^2 \\ y'' & 2 & 6x \end{vmatrix} = 0,$$

because the first column is a linear combination of the final two. Expanding the determinant (by *Laplace's method* or any other legitimate means), we get

$$y \begin{vmatrix} 2x & 3x^2 \\ 2 & 6x \end{vmatrix} - y' \begin{vmatrix} x^2 & x^3 \\ 2 & 6x \end{vmatrix} + y'' \begin{vmatrix} x^2 & x^3 \\ 2x & 3x^2 \end{vmatrix} = 0$$

or

$$x^2y'' - 4xy' + 6y = 0.$$

Thus, knowing the two linearly independent solutions uniquely determines the coefficients $a(x) = -4/x$ and $b(x) = 6/x^2$ in the differential equation satisfied. (Plugging $y_1 = x^2$ and $y_2 = x^3$ into the differential equation would determine $a(x)$ and $b(x)$ even more effectively.) Note that the theory breaks down at $x = 0$, where the coefficients sought become undefined.

Example 4 Given that $y_1(x) = x - 1$ satisfies the equation

$$xy'' + (1 - x)y' + y = 0,$$

since $y_1'' = 0$ and $y_1' = 1$, the general solution for $x \neq 0$ can be obtained by reduction of order. Equivalently, however, Abel's formula implies that the Wronskian

$$W(x) = \begin{vmatrix} y_1 & y \\ y_1' & y' \end{vmatrix} = y_1y' - y_1'y = (x - 1)y' - y$$

of solutions y_1 and y must satisfy

$$W(x) = e^{-\int^x (\frac{1-s}{s})ds}\tilde{c} = \frac{1}{x}e^x c$$

for some constant c. Since $(x - 1)y' - y = \frac{e^x}{x}c$, using the integrating factor $(x - 1)^{-2}$ implies $\left(\frac{y}{x-1}\right)' = \frac{e^x c}{x(x-1)^2}$ and the general solution

$$y(x) = (x - 1)c_1 + (x - 1)\int^x \frac{e^s ds}{s(s - 1)^2}c$$

for some constants c_1 and c and $x \neq 0$.

For the nth-order linear homogeneous equation

$$y^{(n)} + a_1 y^{(n-1)} + \ldots + a_{n-1} y' + a_n y = 0 \qquad (2.11)$$

with constant real coefficients a_i,

$$y(x) = e^{\lambda x} \qquad (2.12)$$

is a solution provided the constant λ satisfies the *characteristic* (or *auxiliary) polynomial*

$$p(\lambda) \equiv \lambda^n + a_1 \lambda^{n-1} + \ldots + a_{n-1}\lambda + a_n = 0. \qquad (2.13)$$

Recall that elementary algebra implies that such monic polynomials (with leading coefficient one) can always be factored as

$$p(\lambda) \equiv (\lambda - \lambda_1)^{m_1}(\lambda - \lambda_2)^{m_2} \ldots (\lambda - \lambda_k)^{m_r} = 0, \qquad (2.14)$$

where the λ_j are distinct roots of $p(\lambda)$, each with its multiplicity $m_j \geq 1$, such that $\sum_{j=1}^{r} m_j = n$. (Admittedly, the actual root-finding is not necessarily simple in practice.) Corresponding to the real roots λ_j of (2.14), then, are real solutions

$$y_j(x) = e^{\lambda_j x}$$

that are defined for all x and that are linearly independent because they have different growth (or decay) rates. Corresponding to complex conjugate roots $\lambda_{j,\ell} = \alpha_j \pm i\beta_j$ of (2.14) are linearly independent complex solutions $\tilde{y}_j(x) = e^{\lambda_j x}$ and $\tilde{y}_\ell(x) = e^{\lambda_\ell x}$, as well as corresponding linearly independent real solutions

$$y_j(x) = e^{\alpha_j x} \cos \beta_j x \quad \text{and} \quad y_\ell(x) = e^{\alpha_j x} \sin \beta_j x,$$

which are the real and imaginary parts (so special linear combinations) of the complex solutions $\tilde{y}_{j,\ell}$ [according to *Euler's* (or *de Moivre's*) *formula* that

$$e^{i\beta} = \cos \beta + i \sin \beta$$

for any real β]. (Readers inexperienced with complex-valued functions are urged to read on, without intimidation. Those knowing the Maclaurin series for the exponential, cosine, and sine series can readily verify Euler's formula. More simply, $z_1 = e^{i\beta}$ and $z_2 = \cos \beta + i \sin \beta$ must be identical since they are both unique solutions of the complex-valued initial value problem $\frac{dz}{d\beta} = iz$, $z(0) = 1$.) We thus have a full set of n linearly independent real solutions to (2.10), whenever the characteristic polynomial (2.14) has n distinct roots. (Indeed, the resulting nonzero Wronskian is then a so-called Vandermonde determinant.) If any root λ_j of (2.14) is, instead, repeated $m_j > 1$ times, there will be m_j corresponding linearly independent (complex) solutions

$$x^q e^{\lambda_j x} \text{ for } q = 0, 1, \ldots, m_j - 1$$

(and their real counterparts), as readers can directly check by expending some labor. In all cases, we readily get n linearly independent solutions $y_i(x)$ to (2.11) (real or complex, as desired) and, thereby, the general solution. We will, primarily, be concerned with second-order differential equations and thereby with characteristic polynomials that are quadratic; the algebra will then be substantially simplified.

Since polynomials of degree two can be explicitly solved, all solutions of the constant coefficient second-order linear homogeneous equation

$$y'' + ay' + by = 0 \tag{2.15}$$

can be found, depending on the nature of the roots

$$\lambda_{1,2} \equiv \frac{1}{2}\left(-a \pm \sqrt{a^2 - 4b}\right)$$

of the corresponding quadratic. The general real solution to (2.15) is therefore given by

$$y(x) = \begin{cases} e^{\lambda_1 x} c_1 + e^{\lambda_2 x} c_2 \text{ if } a^2 > 4b \\ e^{-ax/2}(c_1 + xc_2) \text{ if } a^2 = 4b \\ e^{-ax/2}\left[c_1 \cos\left(x\sqrt{b - \frac{a^2}{4}}\right) + c_2 \sin\left(x\sqrt{b - \frac{a^2}{4}}\right)\right] \text{ if } a^2 < 4b \end{cases} \tag{2.16}$$

for arbitrary constants c_1 and c_2. Note that the ultimate behavior of the solutions (2.16) as $x \to \pm\infty$ depends on the signs of the real parts of the exponents. In particular, if a and b are both positive, the solutions of (2.15) all decay exponentially to zero as $x \to \infty$. This conclusion will, more generally, be referred to as the *asymptotic stability* of the trivial solution of (2.15).

Example 5 Our success in solving linear homogeneous equations allows us to also solve certain nonhomogeneous equations. Consider, for example, the nonhomogeneous initial value problem

$$y'' - y = x, \quad y(0) = y'(0) = 0.$$

If we differentiate the given equation twice, we find that the solution y must also satisfy the homogeneous constant-coefficient fourth-order equation

$$y'''' - y'' = 0,$$

which has the characteristic polynomial $\lambda^4 - \lambda^2 = \lambda^2(\lambda + 1)(\lambda - 1) = 0$. This implies that the y desired must be of the form

$$y(x) = k_1 + xk_2 + e^{-x}c_1 + e^x c_2$$

for constants k_i and c_j. Then, however, $y'' - y = -k_1 - xk_2 = x$ forces us to take $k_1 = 0$ and $k_2 = -1$, so $y(x) = -x + e^{-x}c_1 + e^x c_2$. The initial conditions further require that

$$y(0) = c_1 + c_2 = 0 \quad \text{and} \quad y'(0) = -1 - c_1 + c_2 = 0,$$

so we finally have the unique solution

$$y(x) = -x - \frac{1}{2}e^{-x} + \frac{1}{2}e^x \equiv -x + \sinh x.$$

The procedure used here forms the basis of the (very important) method of *undetermined coefficients*, which we will describe below, along with the related annihilator technique.

Example 6 Consider the *nonhomogeneous* equation

$$y'' + ay' + by = (D^2 + aD + b)y = g(x),$$

where a and b are constants, $g(x)$ is a given continuous forcing function, and $D = \frac{d}{dx}$ is the derivative operator. Factoring this equation as

$$(D - \lambda_1)(D - \lambda_2)y = g(x),$$

using the constants $\lambda_{1,2} \equiv \frac{1}{2}(-a \pm \sqrt{a^2 - 4b})$, which are roots of the characteristic polynomial, let us (somewhat artificially) introduce

$$u \equiv (D - \lambda_2)y = y' - \lambda_2 y$$

as the solution of

$$(D - \lambda_1)u \equiv u' - \lambda_1 u = g(x).$$

(Recall exercise 5 in Chapter 1.) Because we already know how to solve nonhomogeneous first-order equations, we can determine the solution y of the given second-order nonhomogeneous equation by first finding u and then y. Using the integrating factor $e^{-\lambda_1 x}$, we readily get

$$u(x) = e^{\lambda_1 x} c_1 + \int^x e^{\lambda_1(x-t)} g(t) dt$$

and, using the integrating factor $e^{-\lambda_2 x}$,

$$y(x) = e^{\lambda_2 x} c_2 + \int^x e^{\lambda_2(x-r)} u(r) dr.$$

Eliminating $u(r)$ finally yields the desired solution

$$y(x) = e^{\lambda_2 x} c_2 + \left(\int^x e^{\lambda_2(x-r)} e^{\lambda_1 r} dr \right) c_1 + \int^x e^{\lambda_2(x-r)} \int^r e^{\lambda_1(r-t)} g(t) dt dr$$

of the given equation, as can be directly checked by substitution. When $a^2 \neq 4b$, we will, more simply, have

$$y(x) = e^{\lambda_2 x} c_2 + e^{\lambda_1 x} \frac{c_1}{\lambda_1 - \lambda_2} + e^{\lambda_2 x} \int^x e^{(\lambda_1 - \lambda_2) r} \int^r e^{-\lambda_1 t} g(t) dt dr.$$

However, when $a^2 = 4b$, we obtain

$$y(x) = e^{-ax/2} \left(c_2 + x c_1 + \int^x \int^r e^{at/2} g(t) dt dr \right)$$

(as also results by letting $\lambda_1 \to \lambda_2$ in the preceding formula and using l'Hospital's rule). It can be easily verified that the double integral terms represent *particular* solutions of the given nonhomogeneous equation, whereas the first two terms provide the general (*complementary*) solution of the corresponding homogeneous equation. Moreover, the double integrals can be reduced to single integrals by reversing the orders of integration. We thereby obtain the significant conclusion that

$$y(x) = \frac{e^{\lambda_1 x} c_1}{\lambda_1 - \lambda_2} + e^{\lambda_2 x} c_2 + \frac{1}{\lambda_1 - \lambda_2} \int^x (e^{\lambda_1(x-t)} - e^{\lambda_2(x-t)}) g(t) dt \quad \text{for } a^2 \neq 4b$$

and

$$y(x) = e^{-ax/2}(c_2 + x c_1) + \int^x (x-t) e^{-a(x-t)/2} g(t) dt \quad \text{when } \lambda_1 = \lambda_2 = -\frac{a}{2}$$

(using $\int^x \int^r (\cdot) dt dr = \int^x \int_t^x (\cdot) dr dt$ and evaluating the inner integral). Alternatively, we can write the particular solution as

$$\int^x G(x-s) g(s) ds$$

using the Green's function kernel

$$G(z) \equiv \begin{cases} (e^{\lambda_1 z} - e^{\lambda_2 z})/(\lambda_1 - \lambda_2) & \text{if } a^2 \neq 4b \\ ze^{-az/2} & \text{if } \lambda = \lambda_2 = \frac{a}{2}. \end{cases}$$

Because the integral defining the particular solution simplifies considerably when the forcing function $g(x)$ has certain special forms, it will be convenient to develop the method of undetermined coefficients as a quick way to obtain a particular solution.

2.2 Nonhomogeneous Equations

Now consider the nonhomogeneous second-order linear scalar equation

$$L(y) \equiv y'' + a(x)y' + b(x)y = f(x) \tag{2.17}$$

for a nontrivial continuous *forcing function* $f(x)$ and continuous coefficients $a(x)$ and $b(x)$. For convenience, we shall use the differential operator notation

$$L = L(D) \equiv D^2 + a(x)D + b(x),$$

where $D = \frac{d}{dx}$. If y_p is a solution of (2.17), linearity implies that $y_c \equiv y - y_p$ necessarily satisfies the homogeneous equation

$$L(y_c) = L(y) - L(y_p) = f(x) - f(x) = 0,$$

so y_c must be of the form

$$y_c(x) = y_1(x)c_1 + y_2(x)c_2 \tag{2.18}$$

for constants c_1 and c_2 and any pair of linearly independent solutions y_1 and y_2 of the homogeneous equation $L(y) = 0$. Thus, any solution of (2.17) must have the form

$$y(x) = y_1(x)c_1 + y_2(x)c_2 + y_p(x). \tag{2.19}$$

It is standard to call y_p a *particular* solution; y_c, for arbitrary constants c_1 and c_2, the *complementary* solution; and $y(x)$ the *general* solution of (2.17). Algebraists would describe the structure of the resulting set of solutions as an *affine space.*

Our preceding analysis shows that we could seek a particular solution of the nonhomogeneous equation (2.17) in the form

$$y(x) = y_c(x)w(x),$$

where $y_c(x)$ is any nontrivial solution of the homogeneous $L(y_c) = 0$ and $w(x)$ is a variable function which must be determined. We will follow a more traditional path, however. Suppose we know two linearly independent solutions y_1 and y_2 of $L(y) = 0$. The general solution of the homogeneous equation $L(y) = 0$ will be

$$y_c(x) = y_1(x)c_1 + y_2(x)c_2.$$

Moreover, it will also satisfy

$$y_c'(x) = y_1'(x)c_1 + y_2'(x)c_2.$$

The method of *variation of parameters* (sometimes attributed to Lagrange) formally replaces the constants c_1 and c_2 in these two relations by variable functions $u(x)$ and $v(x)$ and seeks a solution $y(x)$ of the nonhomogeneous equation $L(y) = f$ that satisfies both

$$\begin{cases} y(x) = y_1(x)u(x) + y_2(x)v(x) \\ \text{and} \\ y'(x) = y_1'(x)u(x) + y_2'(x)v(x). \end{cases} \tag{2.20}$$

(The method is, naturally, often called *variation of constants*.) Note that the second requirement forces u' and v' to satisfy

$$y_1(x)u'(x) + y_2(x)v'(x) = 0 \tag{2.21}$$

because the two-term formula for the derivative of the products $(y_1u)'$ and $(y_2v)'$ cannot be violated. The differential equation $L(y) = f$ likewise requires that they satisfy

$$y_1'(x)u'(x) + y_2'(x)v'(x) = f(x) \tag{2.22}$$

since

$$\begin{aligned} f(x) = y'' + a(x)y' + b(x)y = (y''{}_1u + y_1'u' + y''{}_2v + y_2'v') \\ + a(x)(y_1'u + y_2'v) + b(x)(y_1u + y_2v) = y_1'u' + y_2'v' \end{aligned}$$

follows since y_1 and y_2 both satisfy $L(y_i) = 0$. Using Cramer's rule to solve the two simultaneous linear equations (2.21) and (2.22) for u' and v', we uniquely obtain

$$(u', v') = \left(\begin{vmatrix} 0 & y_2 \\ f & y_2' \end{vmatrix}, \begin{vmatrix} y_1 & 0 \\ y_1' & f \end{vmatrix} \right) \Big/ \begin{vmatrix} y_1 & y_2 \\ y_1' & y_2' \end{vmatrix},$$

since the denominators are nonzero by the Wronskian test because y_1 and y_2 are linearly independent. Rewriting this as

$$u'(x) = \frac{-f(x)y_2(x)}{W[y_1, y_2](x)} \quad \text{and} \quad v'(x) = \frac{f(x)y_1(x)}{W[y_1, y_2](x)},$$

we integrate to get u and v and, thereby, the general solution

$$y(x) = y_1(x)c_1 + y_2(x)c_2 - y_1(x)\int^x \frac{y_2(t)f(t)dt}{W[y_1, y_2](t)} + y_2(x)\int^x \frac{y_1(t)f(t)dt}{W[y_1, y_2](t)} \tag{2.23}$$

for arbitrary constants c_1 and c_2 (already implied by the unspecified lower limits of integration on the two integrals). More compactly, we identify the last two terms as a particular solution

$$y_p(x) = \int^x G(x, t)f(t)dt, \tag{2.24}$$

using the *Green's function kernel*

$$G(x, t) \equiv \frac{y_2(x)y_1(t) - y_1(x)y_2(t)}{W[y_1, y_2](t)} \equiv \begin{vmatrix} y_1(t) & y_2(t) \\ y_1(x) & y_2(x) \end{vmatrix} \Big/ W(t).$$

The correctness of (2.24) can be checked directly by substituting into the given nonhomogeneous equation (2.17); thus, the seeming arbitrariness of the procedure as it was developed need not be justified step-by-step.

Example 5 (reconsidered) Since $y'' - y = 0$ has linearly independent solutions $e^{\pm x}$, we shall employ variation of constants and seek a solution of the nonhomogeneous equation $y'' - y = x$ such that

$$y = e^x u + e^{-x} v \quad \text{and} \quad y' = e^x u - e^{-x} v.$$

Then,

$$y'' - y = (e^x u + e^{-x} v + e^x u' - e^{-x} v') - (e^x u + e^{-x} v) = x,$$

or

$$e^x u' - e^{-x} v' = x.$$

Likewise, differentiating y and comparing with y', we need

$$e^x u' + e^{-x} v' = 0.$$

The solutions

$$u' = \frac{x}{2}e^{-x} \quad \text{and} \quad v' = -\frac{x}{2}e^{x}$$

of these two linear equations for u' and v' provide us, upon integration,

$$u(x) = \frac{1}{2}\int^{x} te^{-t}dt = -\frac{1}{2}e^{-x}(x+1) + c_1$$

and

$$v(x) = -\frac{1}{2}\int^{x} te^{t}dt = \frac{1}{2}e^{x}(-x+1) + c_2,$$

which gives us the general solution

$$y(x) = -\frac{1}{2}(x+1) + c_1e^{x} + \frac{1}{2}(-x+1) + c_2e^{-x} = c_1e^{x} + c_2e^{-x} - x.$$

We might have, much more easily, guessed the particular solution $y_p(x) = -x$ and simply added the complementary solution to it to get the general solution y. (There is good sense in being adventurous, taking such judicious guesses, and saving lots of tedious work.)

Example 7 The third-order nonhomogeneous equation

$$y''' - 6y'' + 11y' - 6y = f(x)$$

has the complementary solution

$$y_c(x) = e^{x}c_1 + e^{2x}c_2 + e^{3x}c_3,$$

so we might seek a solution of the given nonhomogeneous equation such that

$$\begin{cases} y(x) = e^{x}u(x) + e^{2x}v(x) + e^{3x}w(x) \\ y'(x) = e^{x}u(x) + 2e^{2x}v(x) + 3e^{3x}w(x) \\ \text{and} \\ y''(x) = e^{x}u(x) + 4e^{2x}v(x) + 9e^{3x}w(x). \end{cases}$$

Differentiating y and y' and comparing the last two equations requires u', v', and w' to satisfy both

$$e^{x}u'(x) + e^{2x}v'(x) + e^{3x}w'(x) = 0$$

and

$$e^{x}u'(x) + 2e^{2x}v'(x) + 3e^{3x}w'(x) = 0$$

since y' and y'' would otherwise have additional terms. Substitution for y and its first three derivatives into the given differential equation, after cancellations, requires that

$$e^x u'(x) + 4e^{2x} v'(x) + 9e^{3x} w'(x) = f(x).$$

Solving the three linear equations for u', v', and w' (which have the Wronskian $W(e^x, e^{2x}, e^{3x})$ as their nonzero determinant of coefficients), we uniquely obtain the unknowns

$$u' = \frac{1}{2}e^{-x} f(x), \; v' = -e^{-2x} f(x), \quad \text{and} \quad w' = \frac{1}{2}e^{-3x} f(x),$$

so, upon integrating, we get the general solution

$$y(x) = y_c(x) + \frac{1}{2}\int^x (e^{x-s} - 2e^{2(x-s)} + e^{3(x-s)}) f(s)ds$$

as the sum of the complementary and a particular solution given by the integral of a Green's function times the forcing term f.

All second-order equations of the form

$$L(D)y(x) \equiv y'' + ay' + by = g(x) \tag{2.25}$$

for constants a and b can be solved by the *factorization method* of *example 6*, as well as by variation of parameters. It is more traditional, and perhaps more convenient, however, to solve them by the *method of undetermined coefficients* whenever the function $g(x)$ happens to be a linear combination of polynomials, polynomials times exponentials, or polynomials times exponentials times a sine or a cosine function. (Using complex arithmetic, we will allow g to have the form

$$g(x) = \sum_{j=1}^{m} r_j(x)e^{\omega_j x},$$

where m is finite, the ω_js are (possibly) complex constants, and the r_js are polynomials in x.) Then, we will find a (constant coefficient) annihilating polynomial, or *annihilator*, Q such that the corresponding differential operator $Q(D)$, for $D = \frac{d}{dx}$, *annihilates* g, i.e.,

$$Q(D)g(x) = 0. \tag{2.26}$$

It will take some experience to efficiently select an annihilator corresponding to a given forcing g. For example, if $g_1(x)$ is a polynomial in x of degree $p_1 \geq 0$ such as $x^{p_1} + 5$, $Q_1(D) \equiv D^{p_1+1}$ will satisfy $Q_1(D)g_1(x) = 0$ (because

$D^{n+1}(x^n) = 0$ for any integer $n \geq 0$). If $g_2(x) = r_2(x)e^{\gamma x}$ for a polynomial r_2 of degree p_2, it likewise follows that $Q_2(D) = (D - \gamma)^{p_2+1}$ will satisfy $Q_2(D)g_2(x) = 0$. Finally, if $g_3(x) = s_3(x)e^{\alpha x} \cos \beta x$ or $s_3(x)e^{\alpha x} \sin \beta x$, where s_3 is a polynomial of degree p_3, $Q_3(D) \equiv [(D - \alpha)^2 + \beta^2]^{P_3+1}$ will satisfy $Q_3(D)g_3(x) = 0$. (To convince yourself, try a few examples!) Moreover, any linear combination $g(x) = \delta_1 g_1(x) + \delta_2 g_2(x) + \delta_3 g_3(x)$ will then be annihilated by the product $Q(D) = D^{p_1+1}(D - \gamma)^{p_2+1}[(D - \alpha)^2 + \beta^2]^{p_3+1}$. One cannot, however, similarly find a constant-coefficient annihilator Q so that, for example, $Q(D) \tan x = 0$, but recall that the class of functions $g(x)$ in (2.25) was restricted and $\tan x$, ln x, $1/x$, etc. were not allowed. The method of undetermined coefficients has its limits; it only applies for constant-coefficient operators L and for the special class of g functions described, whereas variation of parameters is always applicable. The annihilator method is just one convenient method to select the form of solution to try when applying the method of undetermined coefficients.

If we operate on the nonhomogeneous equation (2.25) by an annihilator $Q(D)$, (2.26) implies that the solution y will also satisfy the higher-order constant-coefficient homogeneous differential equation

$$Q(D)L(D)y = Q(D)(D^2 + aD + b)y(x) = 0. \tag{2.27}$$

If we presume that $Q(D)$ is a polynomial in D of degree q, any solution y of the $(q + 2)$ nd order differential equation (2.27) must be a linear combination of any $(q + 2)$ of its linearly independent solutions $y_i(x)$. Moreover, we can take the first two, $y_1(x)$ and $y_2(x)$, to be solutions of the second-order homogeneous equation $L(D)y = 0$ (as in (2.16)). It then follows that the particular solution y_p of (2.25) is in the form

$$y_p(x) = y_3(x)k_1 + y_4(x)k_2 + \ldots + y_{q+2}(x)k_q \tag{2.28}$$

for undetermined constant coefficients, $k_1, k_2, \ldots, k_q$. Substituting (2.28) into (2.25) will uniquely determine the unknown k_is. We will illustrate the technique by examples. (Note that we could have added parts of the complementary solution to (2.28), but, being parsimonious, we seek to save unnecessary effort.) Note that a less formal (but completely satisfactory) approach, somewhat different than the annihilator technique, is suggested by exercise 24.

Example 5 (reconsidered) If y satisfies

$$y'' - y = L(D)y \equiv (D^2 - 1)y = x,$$

$D^2 x = 0$ suggests using the annihilator $Q(D) = D^2$ for x to obtain the homogeneous equation

$$D^2(D-1)(D+1)y = 0$$

for y. Thus y must be of the form

$$y(x) = c_1 e^x + c_2 e^{-x} + k_1 + k_2 x$$

and we can seek a particular solution as

$$y_p(x) = k_1 + k_2 x$$

for constants k_1 and k_2 to be determined. However, $(D^2 - 1)(k_1 + k_2 x) = -k_1 - k_2 x = x$ implies that $k_1 = 0$ and $k_2 = -1$, so we obtain the particular solution $y_p(x) = -x$, as we should have guessed earlier.

Example 8 Consider

$$y'' - 3y' - 4y = (D^2 - 3D - 4)y = 3xe^{2x}.$$

Since $(D-2)^2 xe^{2x} = 0$ and $D^2 - 3D - 4 = (D-4)(D+1)$, the solution y also satisfies the homogeneous equation

$$(D-2)^2(D-4)(D+1)y = 0,$$

so it must be of the form $y(x) = e^{4x}c_1 + e^{-x}c_2 + e^{2x}k_1 + xe^{2x}k_2$. Thus, we naturally seek a particular solution

$$y_p(x) = e^{2x}(k_1 + xk_2)$$

for (undetermined) constants k_1 and k_2. Substituting into the given differential equation requires

$$[(D-4)(D+1)][e^{2x}(k_1 + xk_2)] = (-6k_2 x - 6k_1 + k_2)e^{2x} = 3xe^{2x},$$

so both k_1 and k_2 become uniquely specified and we obtain

$$y_p(x) = -\frac{e^{2x}}{12}(1 + 6x).$$

Somewhat more simply, we could have let the exponential forcing $3xe^{2x}$ suggest that

$$y_p(x) = e^{2x} v(x)$$

for some unknown function $v(x)$, The differential equation then implies that v must satisfy

$$v'' + v' - 6v = 3x.$$

Since the forcing $3x$ is linear, we naturally try

$$v(x) = A + Bx$$

and uniquely determine the constants A and B to provide $y_p(x)$ as before.

Example 9 Consider

$$y'' + 4y = (D^2 + 4)y = x(e^x + \sin 2x).$$

Since $(D - 1)^2 xe^x = 0$ and $(D^2 + 4)^2 x \sin 2x = 0$, we take $Q(D) = (D-1)^2(D^2+4)^2$ as the annihilator, so the solution y also satisfies the eighth-order homogeneous equation

$$(D - 1)^2(D^2 + 4)^3 y = 0.$$

Because the corresponding auxiliary polynomial has $\lambda_1 = 1$ as a double root and $\lambda_{2,3} = \pm 2i$ as triple roots, the solution must be of the form

$$\begin{aligned} y(x) &= c_1 \sin 2x + c_2 \cos 2x + e^x(k_1 + k_2 x) \\ &\quad + (\sin 2x)(k_3 x + k_4 x^2) + (\cos 2x)(k_5 x + k_6 x^2). \end{aligned}$$

Since the complementary solution is $y_c(x) = c_1 \sin 2x + c_2 \cos 2x$, we shall seek specific coefficients $k_1, \ldots, k_6$ so that $y_p(x) = y - y_c(x)$ satisfies the given nonhomogeneous equation. Because

$$\begin{aligned} L(D)y_p(x) = (D^2 + 4)y_p(x) &= e^x(5k_2 x + 5k_1 + 2k_2) \\ &\quad + \sin 2x(-8k_6 x + 2k_4 - 4k_5) + \cos 2x(8k_4 x + 4k_3 + 2k_6) \\ &= xe^x + x \sin 2x, \end{aligned}$$

equating coefficients appropriately implies that the six k_js must satisfy the six coupled linear algebraic equations

$$5k_2 = 1, \quad 5k_1 + 2k_2 = 0, \quad -8k_6 = 1, \quad 2k_4 - 4k_5 = 0, \quad 8k_4 = 0, \quad \text{and}$$
$$4k_3 + 2k_6 = 0.$$

Thus, the k_js are uniquely determined and we obtain a particular solution

$$y_p(x) = \frac{1}{25} e^x(-2 + 5x) + \frac{x}{16} \sin 2x - \frac{x^2}{8} \cos 2x.$$

[Observe that just knowing the annihilator Q, and not considering the complementary solution, would suggest the inadequate form $y_p(x) = e^x(k_1 + k_2 x) + (k_3 + k_4 x)\sin 2x + (k_5 + k_6 x)\cos 2x$ for the particular solution. If we, mistakenly, tried it, the procedure would fail, and we would be forced to try the slightly more complicated correct guess.] By contrast, the Green's function solution of example 6 might be simpler than undetermined coefficients (especially since elementary integral is no problem for the software package Maple!).

2.3 Applications

We shall briefly discuss some simple mechanical and electrical *oscillators* that are described by linear second-order equations.

Example 10 Consider a frictionless linear *spring-mass system,* which is *forced* periodically by a cosine function. The displacement $y(t)$ will satisfy an initial value problem for

$$m\ddot{y} + ky = F_0 \cos \omega t$$

on $t \geq 0$. Physically, we can envision a vertically hung spring suspended from a vibrating support, with the displacement measured from the equilibrium position of the unforced system. The mass m and the spring constant k will naturally both be positive; the magnitude F_0 and the frequency ω of the forcing will also be taken to be positive, without loss of generality. Rescaling the coordinates y and t, if necessary, we will simply study the nonhomogeneous operator equation

$$(D^2 + 1)y = \Lambda \cos \omega t, \tag{2.29}$$

for $D = \frac{d}{dt}$, with $y(0)$ and $y'(0)$ prescribed. Since $(D^2 + \omega^2)\cos \omega t = 0$, y will also satisfy the homogeneous equation

$$(D^2 + \omega^2)(D^2 + 1)y = 0.$$

Taking $y_c(t) = c_1 \cos t + c_2 \sin t$ as the complementary solution and

$$y_p(t) = k_1 \cos \omega t + k_2 \sin \omega t$$

as the form of the particular solution, presuming $\omega \neq 1$, we substitute into

$$(D^2 + 1)y_p = k_1(1 - \omega^2)\cos \omega t + k_2(1 - \omega^2)\sin \omega t = \Lambda \cos \omega t$$

to obtain the particular solution

$$y_p(t) = \frac{\Lambda \cos \omega t}{1 - \omega^2}.$$

To satisfy the initial conditions, the general solution $y(t) = c_1 \cos t + c_2 \sin t + \frac{\Lambda \cos \omega t}{1-\omega^2}$ must satisfy $y(0) = c_1 + \frac{\Lambda}{1-\omega^2}$ and $y'(0) = c_2$, so the scaled initial value problem has the unique solution

$$y(t) = y'(0) \sin t + y(0) \cos t + \frac{\Lambda}{1 - \omega^2} (\cos \omega t - \cos t), \tag{2.30}$$

provided $\omega^2 \neq 1$. Note that this solution is periodic when ω is rational. Using the trigonometric identity

$$\frac{\Lambda}{1 - \omega^2}(\cos \omega t - \cos t) = \frac{2\Lambda}{1 - \omega^2} \sin \left[(1 - \omega)\frac{t}{2}\right] \sin \left[(1 + \omega)\frac{t}{2}\right],$$

observe the so-called beat phenomenon as $\omega \to 1$, with the solution for $y(0) = y'(0) = 0$ acting like the periodic *envelope function* $\sin((1 + \omega)t/2)$ with a slowly varying amplitude $\frac{2\Lambda}{1-\omega^2} \sin((1 - \omega)t/2)$, which is also periodic, but on a long time scale. (A plot of this particular solution for ω near 1 displays the situation quite well.)

For $\omega = 1$, the solution

$$y(t) = y'(0) \sin t + y(0) \cos t + \Lambda \frac{t}{2} \sin t \tag{2.31}$$

follows from, instead, either seeking a particular solution

$$y_p(t) = t(k_1 \cos t + k_2 \sin t)$$

(since $(D^2 + 1)^2 y = 0$) or by taking the limit in (2.30) as $\omega \to 1$ using l'Hospital's rule. We call the situation with $\omega = 1$ *resonant*, since the *forcing frequency* $2\pi/\omega$ then coincides with the *natural frequency* 2π of the unforced oscillator. Such resonance makes the solution unbounded (like $t \sin t$ as $t \to \infty$), in contrast to the bounded solution (2.30) for $\omega^2 \neq 1$. One certainly would not expect our simple linear model to continue to apply as the oscillations increased indefinitely in magnitude. The result clearly suggests caution for ω near 1. Readers are encouraged to experience resonance aurally by experimenting with tuning forks.

When Hooke's law does not apply to a given spring, the restoring force ky used for our model will generally need to be replaced by a nonlinear odd function of y, which perhaps also depends on t. A cubic approximation often suffices, for small displacements.

Numerous physical phenomena are attributed to such mechanical resonance, including the effect of Joshua's blast on the walls of Jericho, the critical guideline that troops break step in crossing a bridge, and the collapse of the Tacoma Narrows bridge in heavy winds in 1940. We note that recent work argues that the latter incident should be treated as a nonlinear phenomenon, by taking the suspension cables into account (cf. Lazer and McKenna (1990)).

Example 11 If we apply Kirchhoff's voltage law to the RLC electric circuit shown in Figure 5.

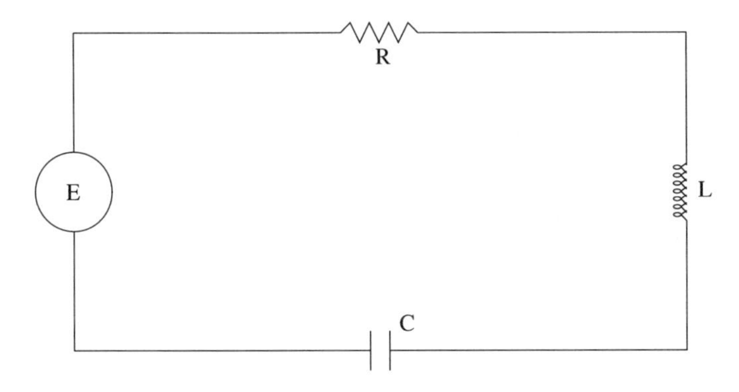

Figure 5: RLC circuit

(presuming "linear" circuit elements), we obtain the voltage balance

$$E(t) = Ri + L\frac{di}{dt} + \frac{Q}{C},$$

where Q is the charge on the capacitor, i is the amount of current flowing, and the positive constants R, L, and C measure resistance, inductance, and capacitance (in appropriate units). Since $i = \dot{Q}$, Q satisfies the second-order linear nonhomogeneous equation

$$L\frac{d^2Q}{dt^2} + R\frac{dQ}{dt} + \frac{Q}{C} = E(t) \tag{2.32}$$

of the special form (2.25) (and the current i satisfies the differentiated equation). Note, however, that more complicated systems arise in analyzing multiloop circuits. We will provide the applied voltage $E(t) = \cos \omega t$, the initial charge $Q(0)$, and the initial current $i(0)$. Using superposition, we could also treat the situation when

$$E(t) = \sum_{k=1}^{\infty} [\alpha_k \cos(k\omega t) + \beta_k \sin(k\omega t)].$$

Here, $Q(t)$ and $i(t)$ will be uniquely determined for all t. (Equation (2.32) can, of course, be simultaneously reinterpreted as providing the deflection of a spring-mass oscillator with a damping force proportional to the velocity. The clear analogy between electrical and mechanical oscillators enables us to simultaneously use our intuition relating to both fields.)

To emphasize the roles of the resistance $R > 0$ and of the forcing frequency $\omega > 0$, we will henceforth take $L = C = 1$ and simply consider the initial value problem

$$\ddot{Q} + R\dot{Q} + Q = \cos \omega t, \;\; Q(0) = q_0, \;\; \dot{Q}(0) = i_0. \tag{2.33}$$

We naturally seek a particular solution as a periodic function of the form

$$Q_p(t) = k_1 \cos \omega t + k_2 \sin \omega t.$$

Letting $D = \frac{d}{dt}$, we get

$$(D^2 + RD + 1)Q_p(t) = [(1 - \omega^2)k_1 + R\omega k_2] \cos \omega t \\ + [(1 - \omega^2)k_2 - R\omega k_1] \sin \omega t = \cos \omega t,$$

and can uniquely obtain k_1 and k_2 and thereby

$$Q_p(t) = \alpha^2[(1 - \omega^2) \cos \omega t + R\omega \sin \omega t],$$

where α is the *amplitude* $[(1 - \omega^2)^2 + R^2\omega^2]^{-1/2}$. Introducing a *phase shift* δ by letting $\cos \delta = (1 - \omega^2)\alpha$ and $\sin \delta = -R\omega\alpha$, we use the formula for $\cos(\omega t + \delta)$ to more conveniently rewrite Q_p as

$$Q_p(t) = \alpha \cos(\omega t + \delta). \tag{2.34}$$

The characteristic polynomial for the homogeneous differential equation corresponding to (2.33) has the roots

$$\lambda_{1,2} = \frac{1}{2}(-R \pm \sqrt{R^2 - 4}),$$

which are both real and negative when $R \geq 2$ and complex conjugates with negative real parts $-R/2$ when $0 < R < 2$. The real complementary solution is therefore

$$Q_c(t) = \begin{cases} c_1 e^{\lambda_1 t} + c_2 e^{\lambda_2 t} & \text{if } R > 2 \\ (c_1 + c_2 t)e^{-t} & \text{if } R = 2 \\ \text{and} & \\ \left[c_1 \cos\left(t\sqrt{1 - \frac{R^2}{4}}\right) + c_2 \sin\left(t\sqrt{1 - \frac{R^2}{4}}\right)\right] e^{-Rt/2} & \text{if } 0 < R < 2 \end{cases}$$

for arbitrary constants c_1 and c_2. (The three cases are respectively called *over-damped*, *critically damped*, and *underdamped*.) c_1 and c_2 will be uniquely determined in each case by the initial conditions so that the sum

$$Q(t) = Q_c(t) + Q_p(t) \tag{2.35}$$

satisfies the initial value problem (2.33). The solution $Q(t)$ decays in all cases to its oscillatory *steady-state* $Q_p(t)$ as $t \to \infty$, since the *transient* part $Q_c(t)$ of the solution decays to zero because $R > 0$.

One might wonder (or experiment to show) how the amplitude α of the steady-state oscillation $Q_p(t)$ changes as ω varies. Since $\frac{d\alpha}{d\omega} = [-2(1-\omega^2) + R^2]\omega\alpha$ can be zero only for $\omega = 0$ or, if $R^2 < 2$, $\omega = \sqrt{1 - R^2/2}$, α will decrease monotonically with ω (from 1) if $R \geq \sqrt{2}$ and will have its maximum $\alpha_{\max} \equiv \left(R\sqrt{1 - \frac{R^2}{4}}\right)^{-1}$ at $\omega = \sqrt{1 - R^2/2}$ if $R^2 < 2$. Thus, for small R, $\alpha_{\max}$ becomes nearly unbounded at this critical frequency (nearly equal to the frequency, 1, of the undamped and unforced oscillator). This possibility parallels the *resonance* we obtained for $\ddot{Q} + Q = \cos t$, as studied for the analogous mechanical system. Other interesting phenomena would occur if we considered the circuit with $R = C = 1$ in (2.32) and let the inductance $L \to 0^+$. The possibilities could be studied experimentally, using readily obtained circuit elements.

2.4 Exercises

1 Find constants a and b so that xe^x satisfies

$$y'' + ay' + by = 0.$$

2 Find values of r so that $y = x^r$ satisfies the differential equation

$$x^2 y'' + axy' + by = 0$$

for given constants a and b and $x \neq 0$.

3 Solve the initial value problem

$$y'' + 2y' + 2y = 0, \ \ y(\pi) = e^{-\pi}, \ y'(\pi) = -2e^{-\pi}.$$

4 Solve the boundary value problem

$$x'' + x' - 6x = 0, \ \ x(0) = 0, \ \ x(p) = 1, \quad \text{for any given } \ p > 0.$$

5 Let $x(t)$ satisfy the initial value problem

$$\ddot{x} + \pi^2 x = b(t) = \begin{cases} \pi^2, & 0 \le t \le 1 \\ 0, & t > 1 \end{cases}$$

with $x(0) = 1$ and $\dot{x}(0) = 0$. Determine the continuously differentiable solution for $t \ge 0$. (Note that it will have a discontinuous second derivative at $t = 1$.)

6 (a) Show that the functions $y_1(x) = x^2$ and $y_2(x) = x|x|$ are linearly independent on $-1 \le x \le 1$.
(b) Compute the Wronskian $W[y_1, y_2](x)$ on the interval.
(c) Explain why y_1 and y_2 cannot both be solutions of a linear homogeneous equation $y'' + a(x)y' + b(x)y = 0$ with smooth coefficients $a(x)$ and $b(x)$ on $-1 \le x \le 1$.

7 Consider the problem

$$y'''' + 8y'' + 16y = 0, \quad \text{with} \quad y^{(j)}(0) \text{ prescribed for } j = 0, 1, 2, \text{ and } 3.$$

For what initial values $y''(0)$ and $y'''(0)$ will solutions be periodic?

8 Solve

$$x^2 y'' - 2xy' + 2y = 6 \ln x$$

for $x > 0$ by either

(a) seeking a solution of the form $y(x) = xv(x)$ with the function v to be determined (since x satisfies the homogeneous equation)

or

(b) solving the transformed problem for $z(t) = y(x)$, where $t = \ln x$ (i.e., $x = e^t$).

9 (a) Suppose $g(x)$ is a given, nonzero solution of

$$g'' + a(x)g' + b(x)g = 0.$$

Find a general solution of

$$y'' + a(x)y' + b(x)y = f(x)$$

for a given forcing function f by seeking a solution in the product form

$$y(x) = g(x)z(x),$$

with the factor $z(x)$ to be determined.

(b) Guess a solution of $y'' + xy' - y = 0$ and use it to find the general solution of

$$y'' + xy' - y = e^{-x^2/2}.$$

10 (a) Suppose $\phi_1(x)$ and $\phi_2(x)$ are two given, linearly independent solutions of a third-order linear equation

$$y''' + a_1(x)y'' + a_2(x)y' + a_3(x)y = 0$$

with given continuous coefficients $a_j(x)$. Find the general solution $y(x)$, using the fact that the equation for $z = y(x)/\phi_1(x)$ will have the solution $\phi_2(x)/\phi_1(x)$, thereby extending the reduction of order method to third-order equations.

(b) Find the general solution of

$$x^5 y''' + x^2 y'' - 2xy' + 2y = 0,$$

given the two linearly independent solutions x and x^2.

11 Use undetermined coefficients to solve the fourth-order equation

$$y'''' + y'' = 3x^2 + 4\sin x - 2\cos x.$$

12 Use variation of parameters to solve

$$y'' + 3y' + 2y = \frac{1}{1+e^x}.$$

13 In *wave propagation*, one sometimes needs to find a complex-valued solution of

$$y'' + \alpha(x)y = 0, \text{ where } \alpha(x) = \begin{cases} 0 & \text{for } 0 < x < 1 \\ 1 & \text{otherwise} \end{cases}$$

in the form

$$y(x) = \begin{cases} e^{ix} + Re^{-ix} & \text{for } x \leq 0 \\ Te^{i(x-1)} & \text{for } x \geq 1 \end{cases}$$

for constants R and T. Use the solution within $0 < x < 1$ to determine a *continuously differentiable* solution on $-\infty < x < \infty$. In particular, find the complex *reflection* and *transmission coefficients* R and T.

14 Solve the initial value problem

$$\dot{x} = Ax + b(t)$$

for

$$x = \begin{pmatrix} x_1 \\ x_2 \end{pmatrix}, \; A = \begin{pmatrix} 4 & 2 \\ 3 & 3 \end{pmatrix}, \; b(t) = \begin{pmatrix} e^t \\ e^{2t} \end{pmatrix}, \; \text{and } x(0) = \begin{pmatrix} 1 \\ 0 \end{pmatrix}$$

by solving the equivalent system of two scalar operator equations

$$(D-4)x_1 = 2x_2 + e^t$$

$$(D-3)x_2 = 3x_1 + e^{2t}$$

for $D = \frac{d}{dt}$. Operate on the first equation by $D-3$, solve the resulting second-order equation for x_1, and then find x_2.

15 Find the general solution of the equation

$$x^2 y'' + x(1+x^2)y' - (1-x^2)y = 2x^3$$

for $x > 0$, given the factorization

$$(D+x)\left(D + \frac{1}{x}\right) y = 2x$$

for $D = \frac{d}{dx}$.

16 Find the general solution of

$$t\ddot{x} - (1+3t)\dot{x} + 3x = 0$$

after first finding a solution of the form

(a) $e^{\lambda t}$ for some constant λ
and
(b) $A + Bt$ for constants A and B.

17 Find the general solution of

$$y'' + xy' + y = 0,$$

given the solution $e^{-x^2/2}$.

18 (a) Seek the solution of the initial value problem

$$(1-t^2)\ddot{y} + 2t\dot{y} - 2y = 0, \quad y(0) = 3, \; \dot{y}(0) = -4$$

as a cubic polynomial $y = At^3 + Bt^2 + Ct + D$.

(b) Solve the initial value problem, given that $y_1 = t$ satisfies the differential equation.

19 Suppose y_1 and y_2 are solutions of $y'' + a(x)y' + b(x)y = 0$. Use the Wronskian to show that y_1 and y_2 are linearly dependent if (a) they both vanish at some point or (b) they both have a maximum or minimum at the same point.

20 Given the solution $y_1(x) = e^{x^2}$ of $y'' - 2xy' - 2y = 0$ and that the *error function*

$$u(x) = \operatorname{erf}(x) \equiv \frac{2}{\sqrt{\pi}} \int_0^x e^{-t^2} dt$$

satisfies $u'' + 2xu' = 0$, find a linearly independent solution $y_2(x)$ of

$$y_2'' - 2xy_2' - 2y_2 = 0,$$

expressed in terms of the error function.

21 Given that $y_1(x) = \sqrt{x}$ satisfies $4x^2y'' + y = 0$, find the general solution of

$$y'' + \frac{y}{4x^2} = \cos x \quad \text{for } x > 0.$$

22 Since e^{ix}, $\cos x$, and $\sin x$ all satisfy the homogeneous equation $y''+y = 0$, e^{ix} must be a linear combination of $\cos x$ and $\sin x$. How does this imply Euler's formula?

23 Use a complex version of variation of parameters to solve

$$y'' + y' + y = x^2.$$

Specifically, set

$$y = e^{\omega x}u(x) + e^{\overline{\omega} x}v(x) \text{ and } y' = \omega e^{\omega x}u + \overline{\omega} e^{\overline{\omega} x}v,$$

where $\omega = e^{2\pi i/3}$ satisfies the auxilary polynomial $\omega^2 + \omega + 1 = 0$. [For a real solution, one can take $v = \overline{u}$ so $y = 2\Re(e^{\omega x}u(x))$, where the complex function $u(x)$ must be determined.]

24 (a) Show that

$$y'' + 2y' + ay = 3 + 4x$$

has a particular solution of the form

$$y(x) = A + Bx \quad \text{when} \quad a \neq 0$$

and of the form

$$y(x) = Ax + Bx^2 \quad \text{when} \quad a = 0$$

(b) Show that

$$y'' + ay' + by = e^x(2 + 3x)$$

has a solution of the form

$$y(x) = e^x v(x),$$

where $v(x)$ is (i) a linear function of x if $a + b + 1 \neq 0$, (ii) a quadratic function if $a + b + 1 = 0$ and $a \neq -2$, and (iii) a cubic function if $a = -2$ and $b = 1$.

(c) Find a complex solution of

$$z'' + 4z = e^{2ix}$$

of the form $z(x) = e^{2ix}w(x)$, where w is a linear function of x, and use the imaginary part of the solution to solve $y'' + 4y = \sin 2x$.

25 Explain why the initial value problem

$$\begin{cases} (x-1)\frac{d^2x}{dt^2} + e^x\frac{dx}{dt} + x = t^2 \\ x(0) = 1,\ \frac{dx}{dt}(0) = 0 \end{cases}$$

cannot have a solution.

26 By considering *Burgers' equation*

$$\ddot{x} + 2x\dot{x} = 0$$

and its linearly independent solutions $x_1(t) = 1$ and $x_2(t) = 1/t$, show that linear combinations of solutions do not necessarily satisfy nonlinear equations.

27 Solve $y'' - y = 2x - 1$.

28 The nonlinear equation

$$M\ddot{y} + F(y) = 0,$$

where

$$F(y) = \begin{cases} K(y+1), & y \leq -1 \\ 0, & |y| < 1 \\ K(y-1), & y \geq 1 \end{cases}$$

for positive constants M and K, can be used to describe a spring with a *dead zone*. Determine the solution corresponding to the initial values

$y(0) = -1$ and $\frac{dy}{dt}(0) = v_0 > 0$ and show that it is periodic with period $T = 2\pi(M/K)^{1/2} + 4/v_0$. (Hint: Patch together a continuously differentiable solution across $y = \pm 1$.)

29 (a) Find a solution of

$$y'' - y = 2f(x)$$

for any smooth function $f(x)$.

(b) Solve the initial value problem

$$y'' - y = 2e^{ax}, \ y(0) = y'(0) = 0 \ \text{ for } a^2 \neq 1.$$

(c) Do the same for $a = \pm 1$.

30 Use the annihilator method to solve the complex-valued equation

$$y'' + 4y = e^{2ix}.$$

31 Use Kirchoff's laws to determine the coupled system of differential equations for the currents I_1 and I_2 passing through the resistors R_1 and R_2 of the circuit pictured.

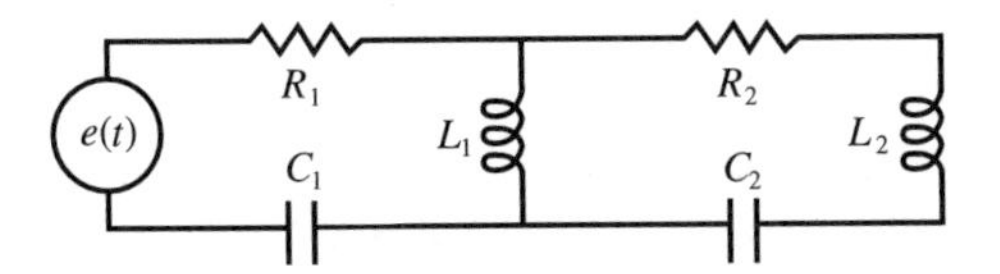

32 (a) The two-point boundary value problem

$$\begin{cases} y'' + \lambda y = 0, & 0 \le x \le 1 \\ y(0) = y(1) = 0 \end{cases}$$

arises in studying the buckling of slender columns. Show that nontrivial solutions occur for loads $\lambda = n^2\pi^2, n = 1, 2, \ldots$.

(b) Find nontrivial solutions of the nonhomogeneous equation

$$\begin{cases} y'' + \pi^2 y = f(x) \\ \text{with} \\ y(0) = y(1) = 0, \end{cases}$$

under some appropriate solvability condition on the function f.

33 (a) Let the linear differential operator L have eigenpairs (λ_k, λ_k) satisfying $Ly_k = \lambda_k y_k$, $k = 1, 2, \ldots, n$. Show that the linear equation

$$Ly + \lambda y = \sum_{k=1}^{n} \alpha_k y_k$$

has solutions

$$y = \sum_{k=1}^{n} \frac{\alpha_k y_k}{\lambda - \lambda_k}, \qquad \text{provided } \lambda \neq \lambda_k.$$

(b) Suppose L is a square matrix. Then, show that the equation

$$(L + \lambda_n)y = \sum_{k=1}^{n} \alpha_k y_k$$

has the solution

$$y = \sum_{k=1}^{n-1} \frac{\alpha_k y_k}{\lambda_n - \lambda_k} + \alpha y_n \quad \text{for an arbitrary constant } \alpha.$$

34 Solve

(a) $y''' = y$,
(b) $y'''' = y + 2e^{2x} + 3e^{x}$,
(c) $y'' = z$, $z'' = y$.

35 (a) Use the transformation $y(x) = \exp(\int^x u(s)ds)$ to convert the second-order linear homogeneous scalar equation

$$y'' + a(x)y' + b(x)y = 0 \tag{L}$$

to the Riccati equation

$$u' + u^2 + a(x)u + b(x) = 0. \tag{R}$$

(b) Use the transformation $u = y'/y$ to convert the Riccati equation (R) to the linear equation (L).
(c) Solve (L) and, thereby, (R) when $b(x) \equiv 0$.

36 Solve the initial value problem

$$y'' + 2y' + 2y = \begin{cases} \frac{1}{k}, & 0 \le t \le k \\ 0, & t > k \end{cases}$$

with $y(0) = y'(0) = 0$ and plot the solution $y(t, k)$ on $0 \le t \le 1$ for a sequence of k values tending to zero. [Idea: Find y and y' at $t = k$ (and their limits as $k \to 0$) by patching y and y' together at $t = k$.]

37 Seek nontrivial approximate solutions $y(x, \epsilon)$ of

$$y'' + \lambda y = \epsilon x^2 y, \ 0 \le x \le \pi, \ y(0) = y(\pi) = 0$$

for ϵ small by obtaining (the first two terms of) formal power series expansions

$$\begin{cases} y(x, \epsilon) = y_0(x) + \epsilon y_1(x) + \epsilon^2 y_2(x) + \dots \\ \lambda(\epsilon) = \lambda_0 + \epsilon\lambda_1 + \epsilon^2\lambda_2 + \dots \end{cases}$$

for the nontrivial eigenfunctions y and the corresponding eigenvalues $\lambda(\epsilon)$.

38 (a) Verify that $e^{\pm t^2}$ satisfies $t\ddot{x} - \dot{x} - 4t^3 x = 0$.

(b) Find a solution of $t\ddot{x} - \dot{x} - 4t^3 x = t^5$ of the form At^2 for some constant A.

(c) Find the general solution of the nonhomogeneous equation in (b).

39 (a) Find solutions of the linear *difference equation*

$$x_{n+2} + x_{n+1} + \frac{1}{4}x_n = 0$$

in the form

$$x_n = c^n$$

for some unknown $c \neq 0$ and $n = 0, 1, 2, \dots$.

(b) What happens to these solutions as $n \to \infty$?

40 (a) Find all solutions of the complex-valued differential equation

$$y'' + (1 + i)y' + iy = 0.$$

(b) Determine all real-valued solutions.
Note: It may be easier to factor (or guess the roots) of the characteristic polynomial than to use the quadratic formula.

41 (a) Show that $y_1(x) = e^{px} \cosh qx$ and $y_2(x) = e^{px} \sinh qx$ are linearly independent solutions of

$$y'' - 2py' + (p^2 - q^2)y = 0.$$

(b) Find more elementary solutions that also span the solution space.

42 Let $y_1(x)$ and $y_2(x)$ be linearly independent solutions of $y'' + a(x)y' + b(x)y = 0$.

(a) Determine the solution $G(x, s)$ of the initial value problem

$$\begin{cases} G'' + a(x)G' + b(x)G = 0 \\ G(s, s) = 0, \ G'(s, s) = 1 \end{cases}$$

for some parameter s, where the prime indicates differentiation with respect to x.

(b) Show that

$$y_p(x) = \int^x G(x,s)f(s)ds$$

satisfies the nonhomogeneous equation

$$y'' + a(x)y' + b(x)y = f(x).$$

43 A spring-mass system consisting of two masses and three springs can be described by the coupled equations

$$\begin{cases} m_1 \frac{d^2u_1}{dt^2} + k_1u_1 + k(u_1 - u_2) = 0 \\ m_2 \frac{d^2u_2}{dt^2} + k_2u_2 + k(u_2 - u_1) = 0 \end{cases}$$

for positive constants m_1, k_1, m_2, k_2, and k.

(a) Show that the system can be reduced to the fourth-order scalar equation

$$(D^2 + \omega_1^2)(D^2 + \omega_2^2)u_j = \nu^4 u_j \text{ for } j = 1 \text{ and } 2$$

when $D = \frac{d}{dt}$, $\omega_j^2 = (k_j + k)/m_j$, and $\nu^4 = k^2/m_1m_2$.

(b) Show that all four roots of the corresponding characteristic polynomial are distinct and purely imaginary and explain why the resulting solutions will be oscillatory. (Hint: $\nu^4 < \omega_1^2\omega_2^2$.)

(c) Assign values to the five constants m_p and k_p. Solve the resulting system numerically and graph the deflections u_1 and u_2 over one period of oscillation.

44 Guess a polynomial solution of

$$(1 + x)y'' + xy' - y = 0$$

and use it to find the general solution, even at the singular point $x = -1$.

45 Show that solutions of

$$y'' - 2ay' + (a^2 + b^2)y = f(x)$$

for constants a and b can be found as

$$y(x) = \int^x G(x - s)f(s)ds$$

for the Green's function $G(r) \equiv \frac{1}{b}e^{-ar}\sinh br$.

46 Consider the two-point boundary value problem

$$y'' + a(x)y' + b(x)y = f(x) \text{ on } 0 \le x \le 1$$

with $y(0)$ and $y(1)$ prescribed. If y_1 and y_2 are linearly independent solutions of the homogeneous equation and $y_p(x)$ is a particular solution of the nonhomogeneous equation, show that the two-point problem will have a unique solution if and only if the determinant

$$\begin{vmatrix} y_1(0) & y_2(0) \\ y_1(1) & y_2(1) \end{vmatrix}$$

is nonzero.

47 (a) Show that the linear equation

$$y'' + a(x)y' + b(x)y = f(x)$$

with $a(x)$ continuously differentiable converts to the simpler equation

$$u'' + c(x)u = d(x),$$

where

$$c(x) \equiv b(x) - \frac{1}{4}a^2(x) - \frac{1}{2}a'(x) \text{ and}$$

$$d(x) \equiv f(x) \exp\left[\frac{1}{2}\int^x a(s)ds\right],$$

under the *Liouville transformation*

$$y(x) = e^{-\frac{1}{2}\int^x a(s)ds}u(x).$$

(b) Show that y has no more zeros than u.

48 (a) Find coefficients $\alpha(x)$ and $\beta(x)$ so that $y_1(x) = x$ and $y_2(x) = e^x$ both satisfy $\alpha(x)y'' + \beta(x)y' + y = 0$.

(b) For these α and β, find a forcing function $f(x)$ so that $y_p(x) = \sin x$ satisfies

$$\alpha(x)y'' + \beta(x)y' + y = f(x).$$

(c) Can variation of parameters be used to solve this nonhomogeneous equation?

49 In polar coordinates, planetary motion is governed by *Kepler's laws*

$$\frac{d^2r}{dt^2} = \frac{c^2}{r^3} + r\rho(t), \quad \text{where } r^2\frac{d\theta}{dt} = c$$

for some constant c and a central force $\rho(t)$ to be prescribed.

(a) Show that $u = 1/r$ satisfies

$$\frac{d^2u}{d\theta^2} + u = -\frac{\rho(t)}{c^2u^3}.$$

(b) In the special case $\rho(t) = -r^{-3}$, show that

$$r^{-1} = 1 + a\cos(\theta - b)$$

for constants a and b.

50 Discuss the possibilities for resonance when

$$y'' + y = \alpha\cos\omega_1 t + \beta\sin\omega_2 t.$$

Solutions to exercises 1–15

1 If $y = xe^x$ satisfies $y'' + ay' + by = 0$, differentiation implies that $(D^2 + aD + b)xe^x = [(2+a) + x(1+a+b)]e^x = 0$, so we need $a = -2$ and $b = 1$. Note that $(D^2 - 2D + 1)e^x = (D-1)^2e^x = 0$ as well.

2 Seeking a solution of $x^2y'' + axy' + by = 0$ in the form $y = x^r$ requires $[r(r-1) + ar + b]x^r = 0$ since $y' = rx^{r-1}$ and $y'' = r(r-1)x^{r-2}$. Thus, $r^2 + (a-1)r + b = 0$ yields the (possibly complex) roots $r = \frac{1-a}{2} \pm \sqrt{\left(\frac{1-a}{2}\right)^2 - b}$.

3 If we seek solutions of $y'' + 2y' + 2y = 0$ in the form $y = e^{\lambda x}$, we will need $\lambda^2 + 2\lambda + 2 = 0$, so $\lambda = -1 \pm i$. The general solution of the differential equation is $y(x) = e^{-x}(c_1\cos x + c_2\sin x)$, so imposing the initial conditions requires $y(\pi) = -e^{-\pi}c_1 = e^{-\pi}$ and $y'(\pi) = e^{-\pi}(-c_2 + c_1) = -2e^{-\pi}$. This uniquely determines the solution $y(x) = e^{-x}(\sin x - \cos x)$.

4 The equation $\ddot{x} + \dot{x} - 6x = 0$ has the general solution $x(t) = e^{-3t}c_1 + e^{2t}c_2$. Thus, the boundary conditions $x(0) = 0$ and $x(p) = 1$ require $c_1 + c_2 = 0$ and $(e^{-3p} - e^{-2p})c_1 = 1$, corresponding to $x(t) = \frac{e^{2t} - e^{-3t}}{e^{2p} - e^{-3p}}$, which is well defined for any $p > 0$.

5 If $x(t)$ satisfies the initial value problem $\ddot{x} + \pi^2x = \pi^2$, $x(0) = 1$, $\dot{x}(0) = 0$ on $0 \le t \le 1$, undetermined coefficients easily provides the unique solution $x(t) \equiv 1$ there. On $t \ge 1$, $\ddot{x} + \pi^2x = 0$, so x must be of the form $x(t) = a\cos\pi(t-1) + b\sin\pi(t-1)$. The continuity conditions $x(1) = 1$ and $\dot{x}(1) = 0$ uniquely determine $x(t) = \cos\pi(t-1)$ for $t \ge 1$. Thus, $\ddot{x}$ and higher even derivatives have a jump discontinuity at $t = 1$.

6 (a) Because x^2 and $x|x|$ are not constant multiples of each other in any open neighborhood of $x = 0$, they are linearly independent there.

(b,c) The Wronskian

$$W(x^2, x|x|) = \begin{vmatrix} x^2 & x|x| \\ 2x & 2|x| \end{vmatrix} = 0$$

everywhere, so the two functions cannot satisfy a linear homogeneous differential equation in standard form, because their trivial Wronskian would then imply their linear dependence.

7 The equation $y'''' + 8y'' + 16y = 0$ has the general solution $y(x) = (c_1 + c_2 x)\cos 2x + (c_3 + c_4 x)\sin 2x$, since $\lambda = \pm 2i$ are both double roots of the characteristic polynomial $(\lambda^2 + 4)^2 = 0$. The only periodic solutions of the differential equations have the form $y(x) = c_1 \cos 2x + c_3 \sin 2x$. Clearly, $c_1 = y(0)$ and $2c_3 = y'(0)$. This, however, requires us to have $y''(0) = -4y(0)$ and $y'''(0) = -4y'(0)$.

8 (a) If $y = xv(x)$ is to satisfy $x^2y'' - 2xy' + 2y = \ln x$ for $x > 0$, v must satisfy $v'' = \frac{6}{x^3}\ln x$. Two integrations yield $v(x) = \frac{3}{x}\ln x + 9/2x + c_1x + c_2$ and $y = xv(x)$.

(b) If $y(x) = z(t)$ for $t = \ln x$, $xy' = \frac{dz}{dt}$ and $x^2y'' = \frac{d^2z}{dt^2} - \frac{dz}{dt}$ convert the equation for $y(x)$ into the equation $\frac{d^2z}{dt^2} - 3\frac{dz}{dt} + 2z = 6t$ for $z(t)$. It has the complementary solution $z_c(t) = c_1e^{2t} + c_2e^t$ and a particular solution $z_p(t) = 3t + 9/2$. The general solution for z implies that for y is therefore $y = c_1x^2 + c_2x + 3\ln x + \frac{9}{2}$.

9 (a) Suppose $g'' + a(x)g' + b(x)g = 0$ and $y'' + a(x)y' + b(x)y = f(x)$. If we set $y = gz$, z will satisfy the first-order linear equation $z'' + \left(\frac{2g'}{g} + a\right)z' = \frac{f}{g}$ for z'. Using the integrating factor $g^2(x)\exp\left[\int^x a(s)ds\right]$, we get $z' = \left[\int^x g(t)f(t)A(t)dt + c_1\right] / \left[g^2(x)\ A(x)\right]$ for $A(x) = \exp\left[\int^x a(s)ds\right]$. Integration of z' then yields

$$y(x) = g(x)\left\{\int^x \frac{1}{g^2(r)A(r)}\left[\int^r g(t)f(t)A(t)dt + c_1\right]dr + c_2\right\}.$$

(b) For $y'' + xy' - y = e^{-x^2/2}$, guess the solution $g(x) = x$ of the homogeneous equation and introduce $A(x) = e^{x^2/2} = \frac{1}{f(x)}$. Following (a), reduction of order implies the general solution $y(x) = \frac{x}{2}\int^x e^{-s^2/2}ds + c_1x\int^x \frac{1}{r^2}e^{-r^2/2}\,dr + c_2x$.

10 (a) If $y''' + a_1(x)y'' + a_2(x)y' + a_3(x)y = 0$ has the solution $\phi_1(x)$, reduction of order implies that $z = y/\phi_1$ will satisfy the second-order equation $z''' + b_1(x)z'' + b_2(x)z' = 0$ for z', where $b_1 \equiv \frac{3\phi_1'}{\phi_1} + a_1$ and $b_2 \equiv \frac{3\phi_1''}{\phi_1} + 2a_1\frac{\phi_1'}{\phi_1} + a_2$ (as the reader can readily verify). Setting $z' = \left(\frac{\phi_2}{\phi_1}\right)' w$ likewise implies that $w'' + c_1(x)w' = 0$ for

$$c_1 \equiv \frac{2\left(\frac{\phi_2}{\phi_1}\right)''}{\left(\frac{\phi_2}{\phi_1}\right)'} + b_1.$$

Thus, $w'(x) = e^{-\int^x c_1(s)ds}$, $z'(x) = \left(\frac{\phi_2(x)}{\phi_1(x)}\right)' \int^x e^{-\int^r c_1(s)ds} dr$, and finally, $y(x) = \phi_1(x) \int^x \frac{d}{dt}\left(\frac{\phi_2(t)}{\phi_1(t)}\right) \int^t e^{-\int^r c_1(s)ds} dr\, dt$.

(b) If we set $y = xz$, then $x^5 y''' + x^2 y'' - 2xy' + 2y = 0$ implies that $x^3 z''' + 3x^2 z'' + z'' = 0$ and, upon integrating, $x^3 z'' + z' = k_1$, a first-order linear equation for z'. Using the integrating factor $\frac{1}{x^3}\exp\left(-\frac{1}{2x^2}\right) = \frac{d}{dx}\left(\exp\left(-\frac{1}{2x^2}\right)\right)$, we get $\exp\left(-\frac{1}{2x^2}\right)z' = k_1 \exp\left(-\frac{1}{2x^2}\right) + k_2$. Solving for z' and integrating again yields $z(x) = k_1 x + k_2 \int^x \exp\left(\frac{1}{2s^2}\right)ds$, so $y(x) = c_1 x + c_2 x^2 + c_3 x \int^x \exp\left(\frac{1}{2s^2}\right)ds$. [(Note: (i) this solution, in contrast to (a), does not explicitly use the given solution x^2, (ii) the given solutions actually satisfy $x^2 y'' - 2xy' + 2y = 0$ and $y''' = 0$, and (iii) a third linearly independent solution is $x \int^x \exp\left[\frac{1}{2s^2}\right] ds$.]

11 If $y'''' + y'' = 3x^2 + 4\sin x - 2\cos x$, two integrations imply that $y'' + y = \frac{x^4}{4} - 4\sin x + 2\cos x + c_1 x + c_2$. (One need not integrate immediately but it will require less work.) The complementary solution is now $y_c(x) = c_3 \cos x + c_4 \sin x$, so we simply seek a particular solution of the form $y_p(x) = Ax^4 + Bx^3 + Cx^2 + \tilde{D}x\cos x + Ex\sin x$, noting that $(D^2+1)(c_1 x + c_2) = c_1 x + c_2$ and $D^2(D^2+1)(c_1 x + c_2) = 0$. Determining $(D^2+1)y_p(x)$ determines the constants A, B, C, and $\tilde{D}$ and the general solution $y(x) = c_1 x + c_2 + c_3\cos x + c_4 \sin x - \frac{1}{4}x^4 - 3x^2 + 2x\cos x + x\sin x$, with arbitrary constants c_i. Note that, by linearity, the polynomial and oscillatory terms in y_p could be separately obtained as responses to their independent forcings.

12 If we seek a solution of $y'' + 3y' + 2y = \frac{1}{1+e^x}$ in the form $y = e^{-2x}u(x) + e^{-x}v(x)$ with $y' = -2e^{-2x}u(x) - e^{-x}v(x)$, we will need $e^{-2x}u' + e^{-x}v' = 0$ and (from the differential equation) $-2e^{-2x}u' - e^{-x}v' = \frac{1}{1+e^x}$. Solving this linear system, $u' = \frac{-e^{2x}}{1+e^x}$ and $v' = \frac{e^x}{1+e^x}$, so integration of $u' = \frac{e^x}{1+e^x} - e^x$

yields $u(x) = \ln(1+e^x) - e^x$ and, likewise, $v(x) = \ln(1+e^x)$, up to additive constants of integration. Hence, $y(x) = c_1e^{-2x} + c_2e^{-x} + [\ln(1+e^x)](e^{-2x} + e^{-x})$.

13 Taking

$$y(x) = \begin{cases} e^{ix} + Re^{-ix} & \text{for } x \le 0 \\ \alpha + \beta x & \text{for } 0 < x < 1 \\ Te^{i(x-1)} & \text{for } x \ge 1, \end{cases}$$

smoothness of y and y' at $x = 0$ and 1 requires $1 + R = \alpha$, $i(1-R) = \beta$, $\alpha + \beta = T$, and $iT = \beta$. Eliminating α and β implies that $T = 1 + R + i(1-R)$ and $T = 1 - R$. Thus, the reflection coefficient is $R = \frac{1}{5}(1-2i)$ and the transmission coefficient is $T = 1 - R$.

14 The vector-matrix system can be rewritten as the two scalar equations $\dot{x}_1 - 4x_1 = (D-4)x_1 = 2x_2 + e^t$ and $\dot{x}_2 - 3x_2 = (D-3)x_2 = 3x_1 + e^{2t}$ for the derivative operator $D = \frac{d}{dt}$. Operating on the first equation by $(D-3)$, we get $(D-3)(D-4)x_1 = 2(D-3)x_2 + (D-3)e^t$. Using the second equation, $(D-3)(D-4)x_1 = 2(3x_1 + e^{2t}) - 2e^t = 6x_1 + 2e^{2t} - 2e^t$. Rearranging, we find that x_1 satisfies $\frac{d^2x_1}{dt^2} - 7\frac{dx_1}{dt} + 6x_1 = (D-6)(D-1)x_1 = 2e^{2t} - 2e^t$. It therefore has the complementary solution $x_{1c} = e^tc_1 + e^{6t}c_2$ and, by undetermined coefficients, the particular solution $x_{1p} = -\frac{1}{2}e^{2t} + \frac{2}{5}te^t$. Taking $x_1(t) = X_{1c} + X_{1p}$ implies that $x_2 = \frac{1}{2}(D-4)x_1 - \frac{1}{2}e^t = -\frac{3}{2}e^tc_1 + e^{6t}c_2 + \frac{1}{2}e^{2t} - \frac{3}{10}e^t - \frac{3t}{5}e^t$. Finally, the initial conditions $x_1(0) = 1$ and $x_2(0) = 0$ specify the solution uniquely with $c_1 = \frac{7}{25}$ and $c_2 = \frac{11}{50}$.

15 If we introduce $u \equiv \left(D + \frac{1}{x}\right)y = y' + \frac{1}{x}y$, the given factorization implies that $(D+x)u = u' + xu = 2x$. Substituting for u and u', we get $y'' + \left(\frac{1}{x} + x\right)y' + \left(1 - \frac{1}{x^2}\right)y = 2x$, so y satisfies the given equation. Thus, if we first integrate $u' + xu = 2x$ to get $u(x) = 2 + c_1e^{-x^2/2}$, and then integrate $xy' + y = xu = 2x + c_1xe^{-x^2/2}$, we get the general solution $y(x) = x - \frac{1}{x}(c_1e^{-x^2/2} + c_2)$.

3

Power series solutions and special functions

3.1 Taylor series

A power series

$$\sum_{n=0}^{\infty} c_n (x - x_0)^n \equiv c_0 + c_1(x - x_0) + c_2(x - x_0)^2 + \dots \tag{3.1}$$

could represent a finite or an infinite sum, depending on whether or not the late coefficients c_n are all zero or not. Since we can always shift the variable from x to $z = x - x_0$, with $z = 0$ corresponding to $x = x_0$, we shall from the beginning suppose that $x_0 = 0$, making equation (3.1) a Maclaurin, rather than a Taylor, series. Using series expansions about nonzero points x_0 may sometimes be preferred in order to retain physical significance, however.

The infinite sum (3.1) with $x_0 = 0$ can converge in some interval $|x| < R$ or it can diverge (which may, nonetheless, provide useful approximate solutions). Readers who have forgotten the needed calculus are urged to review the traditional treatment of convergent series. We can determine the *radius of convergence* R of the series when it exists (from the late terms) through the *ratio test*

$$R = \lim_{n \to \infty} |c_n / c_{n+1}|, \tag{3.2}$$

allowing the possibility that $R \geq 0$ is finite or infinite. We then expect *absolute* convergence of the series

$$\sum_{n=0}^{\infty} c_n x^n = c_0 + c_1 x + c_2 x^2 + \dots \tag{3.3}$$

(i.e., convergence of the majorizing series $\sum_{n=0}^{\infty} |c_n||x|^n)$ within the interval

$-R < x < R$. Note that the formally differentiated series

$$\sum_{n=1}^{\infty} nc_n x^{n-1} = c_1 + 2c_2 x + 3c_3 x^2 + \ldots = \sum_{p=0}^{\infty} (p+1) c_{p+1} x^p$$

will have the same radius of convergence as (3.3). Moreover, the series can be differentiated repeatedly in the same termwise fashion. If an infinitely differentiable function $f(x)$ has the series representation $\sum_{n=0}^{\infty} c_n x^n$ about $x = 0$ (as for analytic functions), we naturally take $c_0 = f(0)$. If $f'(x)$ is further represented by the formally differentiated series, we take $c_1 = f'(0)$ and, continuing, we identify the convergent Maclaurin expansion

$$f(x) = \sum_{n=0}^{\infty} \frac{f^{(n)}(0)}{n!} x^n \tag{3.4}$$

for $|x| < R$. The most common series include

$$e^x = \sum_{n=0}^{\infty} \frac{x^n}{n!}$$

and

$$\sin x = \sum_{n=0}^{\infty} \frac{(-1)^n x^{2n+1}}{(2n+1)!},$$

which converge for all x, and the series

$$\frac{1}{1-x} = \sum_{n=0}^{\infty} x^n,$$

which converges for $|x| < 1$ and is clearly undefined at $x = \pm 1$. Extension of real series to complex arguments x is often a convenient way of learning about such functions and their singularities. Finite Taylor polynomial approximations are, themselves, of much value, especially when the error involved can be explicitly estimated. Note that they share the initial derivatives with the given function f. For *divergent series*, one typically seeks a good approximation of the desired function for, say, small x, by using only a few terms of the series. As x gets smaller, such approximations actually improve. Approximations provided by such divergent *asymptotic* power series often prove to be superior to convergent series for practical computation [see, e.g., Olver (1974)], although they have been seldom discussed in early coursework.

It is natural to seek a power series representation

$$y(x) = \sum_{n=0}^{\infty} c_n x^n \tag{3.5}$$

of solutions to a differential equation

$$f(x, y, y', \ldots, y^{(n)}) = 0 \tag{3.6}$$

about $x = 0$. Termwise differentiation of (3.5) would yield

$$y'(x) = c_1 + 2c_2 x + 3c_3 x^2 + \ldots = \sum_{k=0}^{\infty} (k+1) c_{k+1} x^k,$$

$$y''(x) = 2c_2 + 6c_3 x + \ldots = \sum_{n=2}^{\infty} n(n-1) c_n x^{n-2}$$

$$= \sum_{k=0}^{\infty} (k+2)(k+1) c_{k+2} x^k,$$

etc., so, in general,

$$y^{(j)}(x) = \sum_{k=0}^{\infty} \frac{(k+j)!}{k!} c_{k+j} x^k$$

for any integer $j \geq 0$. Expanding all expressions in the differential equation (3.6) in their own Maclaurin series ultimately provides an expansion

$$f(x, y, y', \ldots, y^{(n)}) = p_0 + p_1 x + p_2 x^2 + \ldots = 0 \tag{3.7}$$

for f with coefficients p_j determined by the c_ns and the Taylor coefficients for f itself expanded about $(0, c_0, c_1, \ldots, n!c_n)$. Successively equating the p_js in (3.7) to zero will, hopefully, termwise determine the coefficients c_n of the proposed series solution (3.5) for y. We will demonstrate this approach through many examples, noting that the laborious process can often be carried out using Mathematica, Maple, and other software to do the often complicated algebra efficiently.

Example 1 Let us seek the series solution $y(x) = \sum_{n=0}^{\infty} c_n x^n$ of the initial value problem

$$y' - y = 1, \quad y(0) = 0.$$

Note that $y(0) = c_0 = 0$ and that the differential equation immediately implies that $y'(0) = c_1 = y(0) + 1 = 1$. Further, differentiating the equation yields

$y'' - y' = 0$, so $y''(0) = 2c_2 = y'(0) = 1$ shows that the series begins as

$$y(x) = x + \frac{x^2}{2} + \ldots.$$

The process can be continued indefinitely, given enough patience.

More formally, but no less correctly, we can substitute the two series for $y(x)$ and $y'(x)$ into the differential equation to obtain

$$\sum_{k=0}^{\infty}(k+1)c_{k+1}x^k - \sum_{n=0}^{\infty} c_n x^n = 1.$$

Remember that the summation index is a "dummy," so we can rewrite this sum as

$$\sum_{j=0}^{\infty}[(j+1)c_{j+1} - c_j]x^j = 1 + \sum_{j=1}^{\infty} 0x^j = 1.$$

Then, equating coefficients successively, we get

$$c_1 - c_0 = 1 \quad \text{and} \quad (j+1)c_{j+1} - c_j = 0 \quad \text{for each } j \geq 1.$$

Since $c_0 = y(0) = 0$, we uniquely obtain $c_1 = 1$ and $c_{j+1} = c_j/(j+1)$ for each $j \geq 1$, so $c_2 = \frac{1}{2}$, $c_3 = \frac{1}{2\cdot 3}, \ldots$, and $c_{j+1} = \frac{1}{(j+1)!}$. Thus,

$$y(x) = \sum_{n=1}^{\infty} \frac{x^j}{j!}.$$

We recognize this as the universally convergent expansion for the unique solution $y(x) = e^x - 1$ of the given initial value problem. Indeed, there was no need to seek a series solution, because the given linear equation can be directly integrated.

Example 2 Let us now seek the general solution of the variable-coefficient linear equation

$$y'' + xy' + y = 0$$

using the power series expansion $y(x) = \sum_{n=0}^{\infty} c_n x^n$ about $x = 0$. Here, $y'(x) = \sum_{n=1}^{\infty} nc_n x^{n-1}$ implies that $xy'(x) = \sum_{n=1}^{\infty} nc_n x^n$, and $y''(x) = \sum_{n=2}^{\infty} n(n-1)c_n x^{n-2} = \sum_{m=0}^{\infty}(m+2)(m+1)c_{m+2}x^m$. (Note that we have conveniently omitted the early terms with zero coefficients.) Using the differential equation, we must have

$$\sum_{m=0}^{\infty}(m+2)(m+1)c_{m+2}x^m + \sum_{n=1}^{\infty} nc_n x^n + \sum_{n=0}^{\infty} c_n x^n = 0$$

and, since the summation index is a dummy variable,

$$2c_2 + c_0 + \sum_{m=1}^{\infty}(m+1)[(m+2)c_{m+2} + c_m]x^m = 0.$$

Equating coefficients, we get $c_2 = -c_0/2$ and, in general, $c_{m+2} = -\frac{c_m}{m+2}$ for each $m \geq 0$. Thus, $c_3 = -\frac{c_1}{3}$, $c_4 = -\frac{c_2}{4} = \frac{c_0}{8}$, $c_5 = -\frac{c_3}{5} = \frac{c_1}{3\cdot 5}$, $c_6 = -\frac{c_4}{6} = -\frac{c_0}{48}$, etc. Here, c_0 and c_1 remain free, corresponding to the unspecified initial values $y(0)$ and $y'(0)$. Rewriting our solution in the suggestive form

$$\begin{aligned} y(x) = y(0)&\left[1 - \frac{x^2}{2} + \frac{x^4}{2^2 2!} - \frac{x^6}{2^3 3!} + \ldots\right] \\ &+ y'(0)\left[x - \frac{2}{3!}x^3 + \frac{2^3}{5!}x^5 - \ldots\right], \end{aligned}$$

we recognize (acknowledging some luck) the first sum as the convergent expansion for

$$y_1(x) = e^{-x^2/2},$$

which we immediately check satisfies the initial value problem $y_1'' + xy_1' + y_1 = 0$, $y_1(0) = 1$, $y_1'(0) = 0$. By reduction of order, we can find a second, linearly independent solution

$$y_2(x) = e^{-x^2/2}\int_0^x e^{s^2} e^{-\int_0^s t\,dt}\,ds = e^{-x^2/2}\int_0^x e^{s^2/2}\,ds$$

of the differential equation. Its Maclaurin expansion can be obtained directly (i.e., by calculating its derivatives at $x = 0$) or by multiplying together the indicated series

$$\begin{aligned} \left(\sum_{j=0}^{\infty}\frac{1}{j!}\left(\frac{-x^2}{2}\right)^j\right)\int_0^x \sum_{k=0}^{\infty}\frac{1}{k!}\left(\frac{s^2}{2}\right)^k ds &= \left(1 - \frac{x^2}{2} + \frac{x^4}{8} - \ldots\right) \\ \times\left(x + \frac{1}{6}x^3 + \frac{x^5}{40} + \ldots\right) &= x - \frac{x^3}{3} + \frac{x^5}{15} + \ldots. \end{aligned}$$

Generally, we will not be fortunate enough to immediately recognize the sums of most series obtained. The closed-form expressions for y_1 and y_2 are certainly convenient for numerically evaluating the solutions for moderate to large values of x; for example, we readily find that $y_1(10)$ is very small, whereas its convergent Maclaurin expansion involves extensive cancellation of large terms with alternating signs. On the other hand, a few terms of this power series give good approximations of, for example, $y_1(1/10)$. Recall, in particular, the convenient error estimate available for series which have alternating signs.

Example 3 We will now seek a series solution $y = \sum_{n=0}^{\infty} c_n x^n$ of the *Airy equation*

$$y'' - xy = 0.$$

Here $y'' = \sum_{k=0}^{\infty}(k+2)(k+1)c_{k+2}x^k$ and $xy = \sum_{k=1}^{\infty} c_{k-1}x^k$, so the equation demands that

$$2c_2 + \sum_{k=1}^{\infty}[(k+2)(k+1)c_{k+2} - c_{k-1}]x^k = 0,$$

i.e., $c_2 = 0$ and $c_{k+2} = \frac{c_{k-1}}{(k+1)(k+2)}$ for all $k \geq 1$, so $c_{n+3} = \frac{c_n}{(n+2)(n+3)}$ for each $n \geq 0$. This leaves c_0 and c_1 free and determines

$$c_3 = \frac{c_0}{2 \cdot 3}, \quad c_6 = \frac{c_3}{5 \cdot 6} = \frac{c_0}{2 \cdot 3 \cdot 5 \cdot 6}, \ldots, c_4 = \frac{c_1}{3 \cdot 4},$$

$$c_7 = \frac{c_4}{6 \cdot 7} = \frac{c_1}{3 \cdot 4 \cdot 6 \cdot 7}, \ldots, c_5 = c_8 = 0, \ldots,$$

and, thereby, the general solution

$$y(x) = y(0)\left[1 + \frac{x^3}{2 \cdot 3} + \frac{x^6}{2 \cdot 3 \cdot 5 \cdot 6} + \ldots \right.$$
$$\left. + \frac{x^{3m}}{2 \cdot 3 \cdot 5 \cdot 6 \cdot \ldots \cdot (3m-1)(3m)} + \ldots \right]$$
$$+ y'(0)x\left[1 + \frac{x^3}{3 \cdot 4} + \frac{x^6}{3 \cdot 4 \cdot 6 \cdot 7} + \ldots \right.$$
$$\left. + \frac{x^{3m}}{3 \cdot 4 \cdot 6 \cdot 7 \cdot \ldots \cdot (3m)(3m+1)} + \ldots \right].$$

These series in x^3 can be shown to converge for all x. They indeed define the *Airy functions* $Ai(x)$ and $Bi(x)$ [cf., e.g., Abramowitz and Stegen (1965)], up to constant multiplicative factors. Such simple linear second-order equations with polynomial coefficients often occur in practical applications.

In general, the power series approach provides convergent series for solutions about *ordinary* or *regular* points. In particular, if $y(x)$ satisfies the linear initial value problem

$$y'' + a(x)y' + b(x)y = f(x), \tag{3.8}$$

with $y(0)$ and $y'(0)$ prescribed, and if the Maclaurin expansions for $a(x)$, $b(x)$, and $f(x)$ converge for $|x| < R$, it is possible to show that the Maclaurin expansion

$$y(x) = \sum_{j=0}^{\infty} \frac{y^{(j)}(0)}{j!} x^j$$

of the solution will also converge for $|x| < R$. Thus, the series solution converges, as long as no singularity of the coefficients in the differential equation limits its convergence. We can readily obtain the terms of the expansion successively, since

$$\begin{aligned} y''(0) = &\ f(0) - a(0)y'(0) - b(0)y(0), \\ y'''(0) = &\ -a(0)y''(0) \\ &\ -a'(0)y'(0) - b(0)y'(0) - b'(0)y(0) + f'(0) \\ = &\ [a^2(0) - a'(0) - b(0)]y'(0) + [a(0)b(0) - b'(0)]y(0) \\ &\ +[f'(0) - a(0)f(0)], \end{aligned}$$

etc. (More traditional series procedures may, however, be more efficient.) Because the coefficients $a(x)$, $b(x)$, and $f(x)$ are smooth at $x = 0$, the point is naturally called a regular point, and the solution $y(x)$ is consequently locally as smooth as the formulas for its derivatives indicate. If the coefficients are only finitely differentiable, the series for y must be appropriately terminated. The variable coefficients in (8) provide no special challenge to obtaining a series solution. In contrast to the linear problem (3.8), we will subsequently only consider nonlinear problems on an ad-hoc basis.

Example 4 The linear nonhomogeneous problem

$$xy' - y = x^2 e^x, \quad y'(0) = 1$$

is complicated because the initial point $x = 0$ is a *singular*, in contrast to a regular, point of the differential equation. If we, nonetheless, attempt to seek a Maclaurin series solution

$$y(x) = \sum_{n=0}^{\infty} c_n x^n, \qquad \text{with } c_1 = 1,$$

we will have $xy'(x) = \sum_{n=1}^{\infty} nc_n x^n$. Since $x^2 e^x = \sum_{m=0}^{\infty} \frac{x^{j+2}}{j!} = \sum_{n=2}^{\infty} \frac{x^n}{(n-2)!}$, the differential equation directly implies that

$$\sum_{n=1}^{\infty} nc_n x^n - \sum_{n=0}^{\infty} c_n x^n = \sum_{n=2}^{\infty} \frac{x^n}{(n-2)!}$$

or

$$-c_0 + (c_1 - c_1)x + \sum_{n=2}^{\infty} \left[(n-1)c_n - \frac{1}{(n-2)!} \right] x^n = 0.$$

Thus, we must have $c_0 = 0$ and $c_n = \frac{1}{(n-1)!}$ for each $n \geq 2$ to uniquely obtain

$$y(x) = x \left(1 + x + \frac{x^2}{2} + \frac{x^3}{6} + \ldots \right).$$

The sum xe^x can be readily verified to be a solution. Even though the usual existence-uniqueness theory does not apply, because the coefficient of y in $y' - \frac{1}{x}y = xe^x$ is undefined at 0, we have determined the unique (fully satisfactory) solution, even at $x = 0$. (The solution also follows by use of the integrating factor $1/x^2$ or by letting $y = xv$ to get a nonsingular equation for v.) The equation can also be solved as a separable equation with $y(0) = 0$ and $y'(0) = \lim_{x \to 0}(y/x) = 1$.

Example 5 For the *nonlinear* equation

$$y' = x^2 + y^2,$$

we can find a unique Maclaurin expansion

$$y(x) = \sum_{j=0}^{\infty} \frac{y^{(j)}(0)}{j!} x^j,$$

where $y(0)$ is arbitrary and higher derivatives $y^{(j)}(0)$ are successively determined (in the usual way) from the differential equation. For the first few coefficients, we expand y' and y^2 to get

$$\begin{aligned}
y' - x^2 - y^2 = &\left(y'(0) + y''(0)x + \frac{1}{2}y'''(0)x^2 + \ldots \right) - x^2 \\
&-[y^2(0) + 2y(0)y'(0)x + (y(0)y''(0) + (y'(0))^2)x^2 + \ldots] \\
= &\ [y'(0) - y^2(0)] + [y''(0) - 2y(0)y'(0)]x \\
&+ \left[\frac{1}{2}y'''(0) - 1 - y(0)y''(0) - (y'(0))^2 \right] x^2 + \ldots = 0.
\end{aligned}$$

Equating terms successively, we now obtain

$$\begin{aligned}
y'(0) = &\ y^2(0),\ y''(0) = 2y(0)y'(0) = 2y^3(0),\ y'''(0) = 2[1 + y(0)y''(0) \\
&+(y'(0))^2] = 2(1 + 3y^4(0)),\ \text{etc.,}
\end{aligned}$$

so

$$y(x) = y(0) + y^2(0)x + y^3(0)x^2 + \left[\frac{1}{3} + y^4(0)\right] x^4 + \dots.$$

No closed-form solution seems obvious; however, the validity of any finite number of terms can be checked by direct substitution into the differential equation.

A curious result is that, for the special initial value $y(0) = 0$, the solution we generated reduces to a power series

$$y(x) = \sum_{j=0}^{\infty} a_j x^{4j+3} = x^3 \left(\frac{1}{3} + \frac{x^4}{63} + \frac{2x^8}{2079} + \dots \right)$$

in powers of x^4. To obtain it directly, note that

$$y'(x) = x^2 \sum_{j=0}^{\infty} (4j+3) a_j x^{4j}$$

and

$$y^2(x) = x^6 \left(\sum_{j=0}^{\infty} a_j x^{4j} \right)^2 = x^6 \sum_{\ell=0}^{\infty} x^{4\ell} \sum_{p=0}^{\ell} a_p a_{\ell-p}$$

(using the *Cauchy formula* for the product of two power series). The differential equation then requires that

$$x^2 \sum_{j=0}^{\infty} (4j+3) a_j x^{4j} = x^2 + x^2 \sum_{j=0}^{\infty} x^{4(j+1)} \sum_{p=0}^{j} a_p a_{j-p}.$$

Thus, $3a_0 = 1$ and

$$(4j+3) a_j = \sum_{p=0}^{j-1} a_p a_{j-1-p} \quad \text{for each } j \geq 1$$

uniquely determines the series termwise.

3.2 Regular singular points and the method of Frobenius

We will now confine our attention to linear homogeneous second-order differential equations

$$y'' + a(x)y' + b(x)y = 0 \tag{3.9}$$

restricted so that either $a(x)$ or $b(x)$ are singular at $x = 0$, but such that both $xa(x)$ and $x^2b(x)$ have convergent power series expansions in an interval about $x = 0$. We will then (apologetically) call $x = 0$ a *regular singular point* of the differential equation (3.9), which we will conveniently rewrite as

$$x^2 y'' + x\alpha(x)y' + \beta(x)y = 0, \tag{3.10}$$

where both $\alpha(x)$ and $\beta(x)$ have convergent Maclaurin expansions about $x = 0$. Moreover, we will seek (possibly singular) solutions to (3.10) in the corresponding *Frobenius form*

$$y(x) = x^r \sum_{j=0}^{\infty} c_j x^j = \sum_{j=0}^{\infty} c_j x^{j+r} \quad \text{with} \quad c_0 \neq 0 \tag{3.11}$$

for $x > 0$. The restriction on c_0 is made so that the power r is unique. Thus, $y(x)$ will behave like a nonzero multiple of x^r for small values of $x > 0$ and r will generally be a complex number, to be determined during the solution process. An analogous expansion can, of course, be generated for $x < 0$ by introducing $s = -x > 0$. For equations with different types of singularities at $x = 0$, we must expect that solutions may be even more singular locally if they exist.

The motivation for seeking solutions like (3.11) comes from studying the special case of a *Cauchy-Euler* (or "equidimensional") *equation*

$$x^2 y'' + \alpha x y' + \beta y = 0,$$

where α and β are real constants. Let us look for solutions of the Cauchy-Euler equation in the form $y(x) = x^\lambda$, for some complex constant λ and $x \neq 0$. Since $y' = \lambda x^{\lambda-1}$ and $y'' = \lambda(\lambda - 1)x^{\lambda-2}$, we will need

$$x^2 y'' + \alpha x y' + \beta y = x^\lambda[\lambda(\lambda - 1) + \alpha\lambda + \beta] = 0,$$

so λ must satisfy the *indicial equation*

$$F(\lambda) \equiv \lambda(\lambda - 1) + \alpha\lambda + \beta = 0, \tag{3.12}$$

i.e., the quadratic $\lambda^2 + (\alpha - 1)\lambda + \beta = (\lambda + \frac{\alpha-1}{2})^2 + \beta - \frac{1}{4}(\alpha - 1)^2 = 0$. The roots

$$\lambda_{1,2} = \frac{1}{2}\left((1 - \alpha) \pm \sqrt{(1 - \alpha)^2 - 4\beta}\right)$$

will be real and distinct if $(\alpha - 1)^2 > 4\beta$, real and equal if $(\alpha - 1)^2 = 4\beta$, and conjugate complex numbers if $(\alpha - 1)^2 < 4\beta$. Thus, the general solution of the Cauchy-Euler equation will have the real form

$$y(x) = x^{\lambda_1} c_1 + x^{\lambda_2} c_2,$$

for $x > 0$ and arbitrary real constants c_1 and c_2, if $(\alpha - 1)^2 > 4\beta$. An analogous representation holds for $x < 0$. Note, in particular, that solutions can be arbitrarily smooth near $x = 0$ if, for example, λ_1 and λ_2 are both positive integers. Generally, however, $x = 0$ is a singularity of the solution y. When $(1 - \alpha)^2 = 4\beta$, we get a nontrivial real solution $y_1(x) = x^{\lambda_1} = x^{(1-\alpha)/2}$, and we can use reduction of order to find a linearly independent real solution as $y_2(x) = y_1 v = x^{\lambda_1} v$, where the differential equation for y forces us to take $xv'' + (2\lambda_1 + \alpha)v' = (xv')' = 0$. This provides y_2 and a general solution for y of the form

$$y(x) = x^{(1-\alpha)/2}(c_1 + c_2 \ln|x|)$$

on both sides of $x = 0$, though with possibly different constants c_1 and c_2 on the two sides. When

$$\lambda_{1,2} = \gamma \pm i\delta \equiv \tfrac{1}{2}(1-\alpha) \pm i\sqrt{\beta - \tfrac{1}{4}(1-\alpha)^2}$$

are conjugate complex, we obtain solutions $x^{\gamma} e^{\pm i\delta \ln|x|}$ and, taking real and imaginary parts, the general real solution

$$y(x) = x^{(1-\alpha)/2}\left[c_1 \cos\left(\ln|x|\sqrt{\beta - \tfrac{1}{4}(1-\alpha)^2}\right)\right.$$
$$\left. + c_2 \sin\left(\ln|x|\sqrt{\beta - \tfrac{1}{4}(1-\alpha)^2}\right)\right]$$

for $x \neq 0$, again with arbitrary real constants c_j.

Example 6 The third-order Cauchy-Euler equation

$$x^3 y''' - x^2 y'' + xy' = 0$$

likewise has solutions $y(x) = x^{\lambda}$ when λ satisfies the corresponding indicial equation

$$[\lambda(\lambda-1)(\lambda-2)] - [\lambda(\lambda-1)] + \lambda = 0$$

or

$$\lambda(\lambda-2)^2 = 0,$$

since $xy' = \lambda x^{\lambda}$, $x^2 y'' = \lambda(\lambda-1)x^{\lambda}$, and $x^3 y''' = \lambda(\lambda-1)(\lambda-2)x^{\lambda}$. This implies the general solution

$$y(x) = c_1 + x^2(c_2 + c_3 \ln|x|)$$

of the differential equation for $x \neq 0$.

Alternatively, we can instead transform the given third-order equation for y into a simpler equation by cleverly introducing

$$z(t) = y(x), \qquad \text{where} \quad x = e^t, \text{ or } t = \ln x, \quad \text{for } x > 0.$$

Under this change of variables,

$$xy' = x\frac{dz}{dt}\frac{dt}{dx} = \frac{dz}{dt},$$

while

$$x^2\frac{d^2y}{dx^2} = x^2\frac{d}{dx}\left(\frac{1}{x}\frac{dz}{dt}\right) = e^{2t}\frac{d}{dt}\left(e^{-t}\frac{dz}{dt}\right)\frac{dt}{dx} = e^t\left(e^{-t}\frac{d^2z}{dt^2} - e^{-t}\frac{dz}{dt}\right)$$
$$= \frac{d^2z}{dt^2} - \frac{dz}{dt} = \frac{d}{dt}\left(\frac{d}{dt} - 1\right)z.$$

Continuing, we find that

$$x^3\frac{d^2y}{dx^3} = \frac{d}{dt}\left(\frac{d}{dt} - 1\right)\left(\frac{d}{dt} - 2\right)z = \frac{d^3z}{dt^3} - 3\frac{d^2z}{dt^2} + 2\frac{dz}{dt},$$

so the given Cauchy-Euler equation for $y(x)$ is transformed to a constant-coefficient differential equation for $z(t)$, namely

$$\left(\frac{d^3z}{dt^3} - 3\frac{d^2z}{dt^2} + 2\frac{dz}{dt}\right) - \left(\frac{d^2z}{dt^2} - \frac{dz}{dt}\right) + \frac{dz}{dt} = 0$$

or

$$\frac{d^3z}{dt^3} - 4\frac{d^2z}{dt^2} + 4\frac{dz}{dt} = 0.$$

Its general solution

$$z(t) = c_1 + e^{2t}(c_2 + c_3t)$$

transforms directly to the solution $y(x)$ already found, and, of course, corresponds to the characteristic polynomial $\lambda^3 - 4\lambda^2 + 4\lambda$ for z that coincides with the indicial equation for y. The transformation $x = -e^q$ could, analogously, be used to obtain a solution for $x < 0$.

Example 7 Consider the two-point problem

$$x^2y'' + xy' + py = 0 \text{ on } 1 \le x \le e^{\pi} \text{ with } y(1) = 0 \text{ and } y(e^{\pi}) = 0$$

and determine which positive constants p correspond to nontrivial solutions y. If we now seek solutions $y = x^{\lambda}$, λ must satisfy the indicial equation

$$\lambda(\lambda - 1) + \lambda + p = \lambda^2 + p = 0,$$

so we obtain complex solutions $x^{\pm i\sqrt{p}} = e^{\pm i\sqrt{p}\ln x} = \cos(\sqrt{p}\ln x) \pm i\sin(\sqrt{p}\ln x)$ of the differential equation. This implies the general real solution

$$y(x) = c_1\cos(\sqrt{p}\ln x) + c_2\sin(\sqrt{p}\ln x)$$

for $x > 0$. Applying the boundary conditions, we find $c_1 = 0$ and $c_2 \sin\sqrt{p}\pi = 0$. Thus, c_2 remains arbitrary, and taking $p = n^2$ for all nonzero integers n, we get the normalized solutions

$$y_n(x) = \sin(n\ln x)$$

with $c_2 = 1$, corresponding to the *eigenvalues* $p = n^2 = 1, 4, \ldots$. For other values of p, only trivial solutions are obtained.

Knowing about Cauchy-Euler solutions, let us again seek solutions to the more general singular differential equation

$$x^2y'' + x\alpha(x)y' + \beta(x)y = 0 \tag{3.10}$$

in the Frobenius form

$$y(x) = \sum_{j=0}^{\infty} c_j x^{j+r} \quad \text{with } c_0 \neq 0. \tag{3.11}$$

Suppose the given series

$$\alpha(x) = \sum_{j=0}^{\infty} \alpha_j x^j \quad \text{and} \quad \beta(x) = \sum_{j=0}^{\infty} \beta_j x^j$$

both converge for $|x| < R$. Since $xy'(x) = x^r\sum_{j=0}^{\infty}(j+r)c_jx^j$ and $x^2y''(x) = x^r\sum_{j=0}^{\infty}(j+r)(j+r-1)c_jx^j$, substitution of the series into (3.10) requires

$$x^r\left[\left(\sum_{j=0}^{\infty}(j+r)(j+r-1)c_jx^j\right) + \left(\sum_{j=0}^{\infty}\alpha_jx^j\right)\left(\sum_{j=0}^{\infty}(j+r)c_jx^j\right) + \left(\sum_{j=0}^{\infty}\beta_jx^j\right)\left(\sum_{j=0}^{\infty}c_jx^j\right)\right] \equiv \sum_{j=0}^{\infty}q_jx^{j+r} = 0. \tag{3.13}$$

Because $c_0x^r \neq 0$ for $x \neq 0$, the leading term requires $q_0 = 0$, so r must satisfy the indicial equation

$$F(r) \equiv r(r-1) + \alpha_0 r + \beta_0 = 0 \tag{3.12}$$

(coinciding with that for the "nearby" Cauchy-Euler equation $x^2y'' + x\alpha_0 y' + \beta_0 y = 0$). For each distinct root

$$r_{1,2} = \frac{1}{2}(1-\alpha_0) \pm \sqrt{\frac{1}{4}(1-\alpha_0)^2 - \beta_0}, \tag{3.14}$$

of F, we seek a corresponding series solution (3.11) of (3.10). Next, considering the coefficient of x^{r+1} in (3.13), we need

$$q_1 = [r(r+1) + \alpha_0(r+1) + \beta_0]c_1 + [\alpha_1 r + \beta_1]c_0 = 0. \tag{3.15}$$

Thus, we can uniquely obtain c_1, for whichever root r is selected, provided the coefficient of c_1 in (3.15) is nonzero. Later coefficients c_j will also then follow successively without complication by setting $q_j = 0$ (as we shall show), at least for that root $r = r_1$ of (3.12) with the larger real part. Moreover, as we might anticipate, the resulting series for y converges for $|x| < R$, except possibly at the regular singular point $x = 0$, i.e., in the "punctured" domain $|x| < R$, $x \neq 0$. When r is a positive integer, the solution obtained is a Maclaurin series whose first $r - 1$ coefficients are all zero.

Note that the *Cauchy product formula* (which invokes rules for multiplying two series) implies that the coefficient of x^{r+j} in (3.13) for each $j > 0$ is

$$\begin{aligned} q_j = {} & [(j+r)(j+r-1) + \alpha_0(j+r) + \beta_0]c_j \\ & + \sum_{k=0}^{j-1} [\alpha_{j-k}(k+r) + \beta_{j-k}]c_k = 0, \end{aligned}$$

or, using the notation of (3.12),

$$F(j+r)c_j = -\sum_{k=0}^{j-1} [(k+r)\alpha_{j-k} + \beta_{j-k}]c_k \quad \text{for each } j \geq 1. \tag{3.16}$$

Recall that the only zeros of $F(z)$ are r_1 and r_2 (cf. (3.14)), with their real parts ordered such that $\Re(r_1 - r_2) \geq 0$. Thus, $F(j + r_1) \neq 0$ for all $j > 0$, and the coefficients c_{j1}/c_{01} in (3.11) are uniquely determined termwise by (3.16) for $r = r_1$. We are thereby guaranteed that we can formally generate one series solution

$$y_1(x) = c_{01} x^{r_1} \sum_{j=0}^{\infty} (c_{j1}/c_{01}) x^j$$

of (3.10), corresponding to the root r_1, with an arbitrary nonzero c_{01}. A second linearly independent solution could then be obtained (using the series representation for y_1) via reduction of order. However, when r_1 and r_2 do not coincide or differ by a (positive) integer, a second solution corresponding to $r = r_2$ can be directly obtained because then $F(j + r_2)$ is also nonzero for all $j \geq 0$ and (3.16) for $r = r_2$ can be uniquely solved termwise for the coefficients c_{j2}/c_{02} in the corresponding series for y_2. We will reconsider the exceptional case below (cf. (3.17)). First, we will present some examples.

Example 8 For

$$2x^2y'' - xy' + (1 + x)y = 0,$$

we will obtain two linearly independent series solutions (3.11). Here, the series for y, y', and y'' and the differential equation require that

$$2\sum_{n=0}^{\infty}(n + r)(n + r - 1)c_n x^{n+r} - \sum_{n=0}^{\infty}(n + r)c_n x^{n+r} + (1 + x)\sum_{n=0}^{\infty} c_n x^{n+r} = 0,$$

or

$$\sum_{n=0}^{\infty}[2(n + r)(n + r - 1) - (n + r) + 1]c_n x^{n+r} + \sum_{n=1}^{\infty} c_{n-1}x^{n+r} = 0,$$

so

$$x^r\left\{(2r^2 - 3r + 1)c_0 + \sum_{n=1}^{\infty}[(2n + 2r - 1)(n + r - 1)c_n + c_{n-1}]x^n\right\} = 0.$$

The resulting indicial equation $F(r) = 2r^2 - 3r + 1 = (2r - 1)(r - 1) = 0$ has roots $r_1 = 1$ and $r_2 = 1/2$. Moreover, later coefficients require that

$$c_{n\,j} = -\frac{c_{n-1\,j}}{(2n + 2r_j - 1)(n + r_j - 1)} \quad \text{for each } n \geq 1 \text{ and } j = 1 \text{ and } 2.$$

The coefficients corresponding to $r_1 = 1$, then, satisfy

$$c_{n\,1} = -\frac{c_{n-1\,1}}{(2n + 1)n},$$

so $c_{11} = -\frac{c_{01}}{1\cdot 3}$, $c_{21} = -\frac{c_{11}}{2\cdot 5} = \frac{c_{01}}{1\cdot 2\cdot 3\cdot 5}$, $c_{31} = -\frac{c_{21}}{3\cdot 7} = -\frac{c_{01}}{1\cdot 2\cdot 3\cdot 3\cdot 5\cdot 7}$, etc. Using induction, we get $c_{m\,1} = \frac{(-1)^m c_{01}}{m!(3\cdot 5\cdot \ldots \cdot(2m+1))}$ and thereby the series solution

$$y_1(x) = c_{01}x\left[1 + \sum_{m=1}^{\infty}\frac{(-1)^m x^m}{m!(3\cdot 5\cdot \ldots \cdot(2m + 1))}\right].$$

Moreover, using the ratio test, $\lim_{m\to\infty}\left|\frac{c_{m1}}{c_{m+1\,1}}\right| = \lim_{m\to\infty}(2m+3)(m+1) = \infty$ implies that the series converges for all x. This is also clear by termwise comparison with the dominating exponential series.

Corresponding to the remaining root $r_2 = 1/2$, we likewise get a second linearly independent solution

$$y_2(x) = c_{02}x^{1/2}\left[1 + \sum_{m=1}^{\infty} \frac{(-1)^m x^m}{m!(1 \cdot 3 \cdot \ldots \cdot (2m-1))}\right],$$

which converges for all $x > 0$, though it loses smoothness at the singular point $x = 0$ due to the $x^{1/2}$ factor. (An analogous pair of linearly independent series solutions could be obtained for $x < 0$.) Taking a linear combination of y_1 and y_2 provides convergent series for all solutions to the given homogeneous equation for $x > 0$.

Setting $y(x) = sz(s)$, however, for $s = \sqrt{x}$ implies that $\frac{d^2z}{ds^2} + 2z = 0$, so we, indeed, simply have

$$y(x) = \sqrt{x}[\alpha \cos\sqrt{2x} + \beta \sin\sqrt{2x}]$$

for arbitrary constants α and β. This makes the series obtained completely obvious since $\sqrt{x}\cos\sqrt{2x}$ will have an expansion in odd powers of $\sqrt{x}$, whereas $\frac{\sin\sqrt{2x}}{\sqrt{x}}$ has an expansion in powers of x.

Example 9 Bessel's equation of order one is defined by

$$x^2y'' + xy' + (x^2 - 1)y = 0.$$

If we set $y(x) = \sum_{n=0}^{\infty} c_n x^{n+r}$, then the differential equation requires that

$$\sum_{n=0}^{\infty} c_n[(n+r)(n+r-1) + (n+r) - 1]x^{n+r} + \sum_{n=0}^{\infty} c_n x^{n+r+2} = 0.$$

Simplifying the coefficient of x^{n+r}, we have

$$[r(r-1) + r - 1]c_0x^r + [(r+1)r + (r+1) - 1]c_1x^{r+1}$$

$$+ \sum_{n=2}^{\infty}[(n+r-1)(n+r+1)c_n + c_{n-2}]x^{n+r} = 0.$$

The roots $r_{1,2} = \pm 1$ of the relevant indicial equation $F(r) = r(r-1)+r-1 = 0$ differ by an integer. Thus, we are only guaranteed one solution of the form (3.11) (corresponding to $r = r_1 = 1$). For that solution, later terms require that

$$3c_{11} = 0 \quad \text{and} \quad r = (n^2 + 2n)c_{n1} + c_{n-21} = 0 \quad \text{for each } n > 1.$$

Thus, the odd terms are all zero and the even terms satisfy $c_{21} = -\frac{c_{01}}{8}$, $c_{41} = -\frac{c_{21}}{24} = \frac{c_{01}}{192}, \ldots$, and so we get the series representation

$$y_1(x) = c_{01}x\left(1 - \frac{1}{2!}\left(\frac{x}{2}\right)^2 + \frac{1}{2!\,3!}\left(\frac{x}{2}\right)^4 - \ldots\right),$$

which is readily seen to converge for all x.

If we tried to obtain a second linearly independent solution of the form (3.11) corresponding to $r = r_2 = -1$, we would need $-c_{12} = 0$ and $(n^2 - 2n)c_{n2} + c_{n-22} = 0$ for each $n \geq 2$. For $n = 2$, however, this would force us to take $c_{02} = 0$, which contradicts our assumption that the leading coefficient in (3.11) is nonzero. This shows that no second linearly independent solution can be represented in the form (3.11) with $r = r_2$. We, instead, use reduction of order to obtain the solution

$$y_2(x) = y_1(x)\int^x \frac{dt}{ty_1^2(t)} \qquad \text{for } x > 0.$$

Formally substituting the power series expansion for y_1 under the integral sign determines

$$\begin{aligned}
y_2(x) &= \frac{y_1(x)}{c_{01}^2}\int^x \frac{dt}{t^3(1 - \frac{t^2}{8} + \frac{t^4}{192} - \ldots)^2} \\
&= \frac{y_1(x)}{c_{01}^2}\int^x \frac{dt}{t^3(1 - \frac{t^2}{4} + \frac{5t^4}{192} - \ldots)} \\
&= \frac{y_1(x)}{c_{01}^2}\int^x \left(\frac{1}{t^3} + \frac{1}{4t} + \frac{7}{192}t + \ldots\right) \\
&= \frac{1}{4}\frac{y_1(x)}{c_{01}^2}\ln x + \frac{x}{c_{01}}\left(1 - \frac{x^2}{8} + \frac{x^4}{192} + \ldots\right)\left(-\frac{1}{2x^2} + \frac{7x^2}{384} + \ldots\right) \\
&= \frac{y_1(x)\ln x}{4c_{01}^2} - \frac{1}{2c_{01}x}\left(1 - \frac{x^2}{8} - \frac{x^4}{32} + \ldots\right).
\end{aligned}$$

Although y_1 is smooth at the singular point, we find that y_2 (like most other solutions) is not, due to the $\frac{1}{x}$ singularity and to the $x^k \ln x$ terms occurring for odd powers $k > 0$.

Example 10 Consider

$$xy'' - y' + 4x^3y = 0.$$

Rewriting the equation as $x^2y'' - xy' + 4x^4y = 0$, it has the form (3.10) with $\alpha(x) = -1$ and $\beta(x) = 4x^4$, so $x = 0$ is a regular singular point. Seeking solutions in the Frobenius form (3.11) clearly requires that

$$c_0 r(r-2) + c_1 x(r+1)(r-1) + c_2 x^2 (r+2) r + c_3 x^3 (r+3)(r+1) + \sum_{m=4}^{\infty} [c_m (r+m)(r+m-2) + 4c_{m-4}] x^m = 0.$$

Since $c_0 \neq 0$, r must satisfy the indicial equation

$$F(r) = r(r-2) = 0,$$

with roots $r_1 = 2$ and $r_2 = 0$ differing by an integer. For $r = 2$, the resulting series

$$3c_{11}x + 8c_{21}x^2 + 15c_{31}x^3 + \sum_{m=4}^{\infty} [(m+2)mc_{m1} + 4c_{m-41}]x^m = 0$$

must be zero termwise, so $c_{11} = c_{21} = c_{31} = 0$ and $c_{m1} = \frac{-4c_{m-41}}{(m+2)m}$ for each $m \geq 4$. This ultimately provides us with the series solution

$$y_1(x) = c_{01} x^2 \left(1 - \frac{x^4}{6} + \frac{x^8}{120} - \frac{x^{12}}{5040} + \dots\right),$$

which one might recognize as the smooth function $y_1(x) = c_{01} \sin x^2$. We could get a second linearly independent solution $\cos x^2$ by guessing, by reduction of order, or by seeking a second Frobenius solution

$$y_2(x) = \sum_{k=0}^{\infty} c_{k2} x^k \quad \text{with} \quad c_{02} \neq 0$$

such that

$$-c_{12}x + 3c_{32}x^3 + \sum_{m=4}^{\infty} [m(m+2)c_{m2} + 4c_{m-42}]x^m = 0.$$

This requires $c_{12} = c_{32} = 0$ and $c_{m2} = \frac{-4c_{m-42}}{m(m+2)}$ for $m \geq 4$. The coefficient c_{22} is free, but we will take it to be zero (otherwise, the y_2 we generate will include a multiple of y_1). We then obtain

$$y_2(x) = c_{02} \left(1 - \frac{x^4}{2} + \frac{x^8}{24} - \dots\right) = c_{02} \cos x^2,$$

and the general solution for *all* x is simply

$$y(x) = c_1 \sin x^2 + c_2 \cos x^2,$$

with arbitrary constants c_1 and c_2. Thus, all solutions are smooth at the singular point $x = 0$. The Frobenius technique works, without difficulty, even though the roots of the indicial equation (3.12) differ by integer values.

A general result for equations with regular singular points is that when the roots of the indicial equation (3.12) differ by an integer, the method of Frobenius always finds one solution $y_1(x)$ of the form (3.11) corresponding to the larger root r_1 of (3.12). It then generally seeks a second linearly independent solution of the (peculiar) form

$$y_2(x) = a y_1(x) \ln|x| + x^{r_2}\left(c_{02} + \sum_{n=1}^{\infty} c_{n2} x^n\right) \quad \text{with } c_{02} \neq 0. \tag{3.17}$$

[Questioning readers might not be comfortable accepting this fact, but we hope experience (or later study) will convince them that it is prudent to rely on Frobenius and proceed.] When we are lucky, as in example 10, the constant a is 0, and the second solution is again of the form (3.11), but now with $r = r_2$. Otherwise, we will simply determine the coefficients c_{n2} in (3.17) for $n \geq 1$ successively, as we shall illustrate below for Bessel's equation of any integer order $p > 0$. First, we will show that we might analogously generate a *divergent* power series, when the equation is not exactly of the specified form (3.10).

Example 11 The linear equation

$$x^2 y'' + (1 + 3x)y' + y = 0$$

has $x = 0$ as an irregular singular point since the differential equation is not of the special form (3.10) with smooth coefficients $\alpha(x)$ and $\beta(x)$ at $x = 0$. (It has the form (3.10), but $\alpha(x) = 3 + 1/x$ blows up at the origin.) We might, nonetheless, still seek a solution as the Frobenius series

$$y(x) = \sum_{n=0}^{\infty} c_n x^{n+r} \quad \text{with } c_0 \neq 0.$$

Substituting into the differential equation and combining coefficients then requires that

$$\sum_{n=0}^{\infty} [(n+r)(n+r-1) + 3(n+r) + 1] c_n x^{n+r} + \sum_{n=0}^{\infty} (n+r) c_n x^{n+r-1} = 0$$

or

$$r c_0 x^{r-1} + \sum_{n=0}^{\infty} [(n+r+1)c_{n+1} + [(n+r)^2 + 2(n+r) + 1]c_n] x^{n+r} = 0.$$

Thus, we take $r = 0$ and $c_{n+1} = -(n+1)c_n$ for each $n \geq 0$, yielding the formal expansion

$$y(x) = y(0) \sum_{n=0}^{\infty} (-1)^n n! x^n.$$

This series is divergent, since its radius of convergence is zero by the ratio test. For small x, however, the first few terms do, indeed, provide a very useful *asymptotic solution* of the differential equation. [See Poincaré (1892).] It remains appropriate to obtain an *asymptotic approximation* for another linearly independent solution, but that is beyond the scope of our present study.

3.3 Bessel functions

Now we introduce Bessel's equation of order p,

$$x^2 y'' + xy' + (x^2 - p^2)y = 0, \tag{3.18}$$

which has $x = 0$ as a regular singular point. We will take p to be an arbitrary nonnegative real constant, although complex p values might be more generally considered. If we seek a solution in the Frobenius form

$$y(x) = \sum_{n=0}^{\infty} c_n x^{n+r} \quad \text{for} \quad c_0 \neq 0 \text{ and } x > 0, \tag{3.11}$$

Bessel's equation requires

$$(r^2 - p^2)c_0 x^r + [(r+1)^2 - p^2]c_1 x^{r+1} + \sum_{n=2}^{\infty} [((r+n)^2 - p^2)c_n + c_{n-2}]x^{r+n} = 0.$$

The indicial equation $r^2 - p^2 = 0$ has the two roots $r_{1,2} = \pm p$, which differ by an integer whenever p is an integral multiple of $1/2$. Whatever p, we are always guaranteed that we can find a solution y_1 using $r_1 = p \geq 0$ in (3.11). It turns out to yield the smooth even series

$$x^{-p} y_1(x) = c_{01} \left[1 + \sum_{m=0}^{\infty} \frac{(-1)^m x^{2m}}{2^{2m} m!(p+1)(p+2)\dots(p+m)} \right],$$

which is easily seen to converge for all x. Because p enters Bessel's equation as p^2, we might expect the series with p replaced by $-p$ to also be a solution. Note, however, that singular coefficients occur when p is taken as a negative integer. It is traditional to compensate for this potential singularity by letting $c_{01} = \frac{1}{2^p \Gamma(p+1)}$, using the gamma function Γ, which we will introduce presently.

We shall define

$$J_p(x) \equiv \left(\frac{x}{2}\right)^p \sum_{n=0}^{\infty} \frac{(-1)^n(\frac{x}{2})^{2n}}{n!\Gamma(p+n+1)} \qquad \text{for } x > 0 \tag{3.19}$$

as the *Bessel function* of order p (of the first kind). These special function solutions of (3.18) are extremely important in describing physical solutions to many applied problems, especially ones involving a cylindrical symmetry.

The *gamma function*, a generalization of the factorial, will be defined for all $p > -1$ by the integral

$$\Gamma(p+1) \equiv \int_0^\infty e^{-t} t^p \, dt. \tag{3.20}$$

If we integrate by parts, we get the *recurrence formula* $\Gamma(p+1) = \frac{\Gamma(p+2)}{p+1}$, or (upon shifting the argument)

$$\Gamma(q+1) = q\Gamma(q),$$

since

$$\Gamma(p+1) = \frac{1}{p+1}\left(e^{-t}t^{p+1}\Big|_0^\infty + \int_0^\infty e^{-t}t^{p+1}dt\right)$$

and the boundary contributions are both trivial. Thus, $\Gamma(p+1) = p(p-1)\ldots(p-j)\Gamma(p-j)$ and, because $\Gamma(1) = \int_0^\infty e^{-t}dt = 1$, it follows that $\Gamma(p+1) = p!$ for any integer $p \geq 0$. Moreover, the recurrence formula shows that $\Gamma(0)$ is not defined, because this would require $0\Gamma(0) = \Gamma(1) = 1$. The formula does, however, allow us to extend the definition of the gamma function to negative, but nonintegral, arguments. The real function $\Gamma(q)$ so defined will be singular only when $q = 0, -1, -2, \ldots$. It is indeed simple to extend the definition of $\Gamma(p)$ as an analytic function of a complex argument p, except only for simple poles at zero and negative integer values of p. We observe that it is often convenient to denote $\Gamma(q+1)$ simply as $q!$, so the *factorial* becomes defined in the full complex plane, with only the countable infinity of singularities encountered.

When $2p$ is not a positive integer, it is clear that

$$J_{-p}(x) \equiv \left(\frac{2}{x}\right)^p \sum_{n=0}^{\infty} \frac{(-1)^n \left(\frac{x}{2}\right)^{2n}}{n!\Gamma(-p+n+1)}$$

defines a second linearly independent solution of (3.18) (since $J_{\pm p}(x)$ behave like multiples of $x^{\pm p}$ near the singular point 0). Taking $1/\Gamma(-p+n+1) = 0$ for $n = 0, 1, \ldots, p-1$ shows that $J_p(x) = (-1)^p J_{-p}(x)$ when p is an integer.

Otherwise, the general solution of (3.18) will actually take the form

$$y(x) = k_1 J_p(x) + k_2 J_{-p}(x) \tag{3.21}$$

for arbitrary constants k_1 and k_2. When $p = 1/2$, for example, we have

$$J_{1/2}(x) = \sqrt{\frac{x}{2}} \sum_{n=0}^{\infty} \frac{(-1)^n \left(\frac{x}{2}\right)^{2n}}{n!\Gamma\left(\frac{3}{2}+n\right)}.$$

Now, $\Gamma\left(\frac{3}{2}\right) = \frac{1}{2}\Gamma\left(\frac{1}{2}\right), \Gamma\left(\frac{5}{2}\right) = \frac{3}{2^2}\Gamma(1/2)$, etc., whereas $\Gamma\left(\frac{1}{2}\right) = 2\int_0^\infty e^{-s^2}\,ds = \sqrt{\pi}$. Thus,

$$J_{\frac{1}{2}}(x) = \sqrt{\frac{x}{2\pi}}\left(2 - \frac{x^2}{3} + \frac{x^4}{60} - \dots\right) = \sqrt{\frac{2}{\pi x}}\sin x,$$

as we can directly check by substituting into Bessel's equation with $p = 1/2$. Analogously, we obtain the linearly independent solution

$$J_{-\frac{1}{2}}(x) = \sqrt{\frac{2}{x}} \sum_{n=0}^{\infty} \frac{(-1)^n (\frac{x}{2})^{2n}}{n!\Gamma\left(n+\frac{1}{2}\right)} = \sqrt{\frac{2}{\pi x}}\cos x$$

of the same equation. This suggests that, instead of directly seeking solutions $y(x)$ to Bessel's equation of order $1/2$, it might be clever to let

$$y(x) = u(x)/\sqrt{x}.$$

Then, $xy' = \frac{1}{2\sqrt{x}}(-u + 2xu')$ and $x^2y'' = \frac{1}{4\sqrt{x}}(3u - 4xu' + 4x^2u'')$ converts this Bessel equation to $x^{3/2}(u'' + u) = 0$, providing us the general (smooth) solution $y(x) = \frac{1}{\sqrt{x}}(c_1\cos x + c_2\sin x)$ on any interval not containing $x = 0$.

When p is zero or a positive integer,

$$\begin{aligned} J_p(x) &= \left(\frac{x}{2}\right)^p \sum_{m=0}^{\infty} \frac{(-1)^m}{m!\Gamma(m+p+1)}\left(\frac{x}{2}\right)^{2m} \\ &= \left(\frac{x}{2}\right)^p \sum_{m=0}^{\infty} \frac{(-1)^m}{m!(m+p)!}\left(\frac{x}{2}\right)^{2m}. \end{aligned}$$

To get a second linearly independent solution of Bessel's equation of order p, we could use reduction of order, as in example (9), or could instead set

$$y_2(x) = a(\ln|x|)J_p(x) + x^{-p}C(x) \text{ with } C(x) = \sum_{k=0}^{\infty} c_k x^k \tag{3.22}$$

for $c_0 \neq 0$ and $x \neq 0$, corresponding to the general Frobenius form (3.17). Carrying out the differentiations and substituting into Bessel's equation implies, after cancellations, that

$$2ax^{p+1}J_p' + x^2C'' + (1-2p)xC' + x^2C = 0.$$

Now, however, we can replace $xJ_p'(x)$ by its series

$$xJ_p'(x) = \sum_{m=0}^{\infty}(2m+p)d_{2m}x^{2m+p}$$

with $d_{2m} \equiv \frac{(-1)^m}{2^{2m+p}m!(m+p)!}$. Leaving c_0 arbitrary, the singular nonhomogeneous differential equation for C requires that its coefficients satisfy

$$(1-2p)c_1x + \sum_{k=2}^{\infty}[k(k-2p)c_k + c_{k-2}]x^k + 2ax^{2p}\sum_{m=0}^{\infty}(2m+p)d_{2m}x^{2m} = 0.$$

We must take $c_1 = 0$ and then we obtain an even power series for C with later coefficients c_k for k even satisfying

$$k(k-2p)c_k + c_{k-2} = 0 \quad \text{for} \quad 2 < k < 2p$$

and

$$k(k-2p)c_k + c_{k-2} + 2a(k-p)d_{k-2p} = 0 \quad \text{for} \quad k \geq 2p.$$

Setting $k = 2j$ and using mathematical induction, we obtain

$$c_{2j} = \frac{c_0}{2^{2j}j!(p-1)(p-2)\cdot\ldots\cdot(p-j)} \text{ for } j = 1, 2, \ldots, p-1. \quad (3.23)$$

Moreover, $0c_{2p} + c_{2p-2} + 2apd_0 = 0$ implies that we must select

$$a = -\frac{c_0}{2^{p-1}(p-1)!} \neq 0. \quad (3.24)$$

Later terms then follow successively as

$$c_{2(p+m)} = \frac{1}{m(p+m)}\left[-\frac{1}{4}c_{2(p+m-1)} + \frac{c_0(p+2m)d_{2m}}{2^p(p-1)!}\right], \quad m = 1, 2, \ldots. \quad (3.25)$$

This formula corresponds to the terms of the previously obtained second solution for Bessel's equation of order 1. Finally, we normalize our solution by selecting c_0 so that $a = 1$ and thereby we define the Bessel function (of the second kind)

$$K_p(x) \equiv -\frac{1}{2}\left(\frac{2}{x}\right)^p \sum_{j=0}^{p-1} \frac{(p-j-1)!}{j!}\left(\frac{x}{2}\right)^{2j} + (\ln|x|)J_p(x)$$

$$-\frac{1}{2p!}\left(1+\frac{1}{2}+\ldots+\frac{1}{p}\right)\left(\frac{x}{2}\right)^p - \frac{1}{2}\left(\frac{x}{2}\right)^p \sum_{m=1}^{\infty} \frac{\left(-x^2/4\right)^m}{m!(m+p)!}$$

$$\left[\left(1+\frac{1}{2}+\ldots+\frac{1}{m}\right)+\left(1+\frac{1}{2}+\ldots+\frac{1}{m+p}\right)\right] \quad (3.26)$$

of integer order $p \geq 1$. The solutions $J_p(x)$ and $K_p(x)$ turn out to be linearly independent.

Despite all our hard work, however, instead of using $K_p(x)$, it is traditional (and, believe it or not, more convenient) to introduce a new Bessel function

$$Y_p(x) \equiv \frac{\cos p\pi\, J_p(x) - J_{-p}(x)}{\sin p\pi}, \quad (3.27)$$

which is defined for *all* values of p. Using limiting arguments (i.e., l'Hospital's rule) for integer values of p, we obtain

$$Y_p(x) = \frac{1}{\pi}\left[\left.\frac{\partial J_\nu(x)}{\partial \nu}\right|_{\nu=p} + (-1)^p \left.\frac{\partial J_\nu(x)}{\partial \nu}\right|_{\nu=-p}\right]$$

(cf. Olver (1974)). In general, $x^2 J_p'' + xJ_p' + (x^2 - p^2)J_p = 0$ and $x^2 Y_p'' + xY_p' + (x^2 - p^2)Y_p = 0$. Multiplying the first equation by Y_p, the second by J_p, and subtracting, we obtain $-x^2(J_p Y_p'' - Y_p J_p'') - x(J_p Y_p' - Y_p J_p') = 0$, or equivalently, $[x(J_p Y_p' - Y_p J_p')]' = 0$ for $x \neq 0$. (We admit that these time-honed manipulations are not ones that novices could readily think up.) Introducing the Wronskian W, this last result implies that

$$xW[J_p, Y_p](x) = \text{ constant.}$$

However, $xW[J_p, Y_p](x) = -\frac{x}{\sin \pi p} W[J_p, J_{-p}](x)$, by linearity, since $W[J_p, J_p] = 0$. We can use the leading terms of our power series for J_p and J_{-p} to determine the constant

$$xW[J_p, Y_p](x) = \lim_{x \to 0}\left(-\frac{x}{\sin \pi p}\begin{vmatrix} \dfrac{x^p}{\Gamma(p+1)2^p} & \dfrac{2^p}{x^p\Gamma(1-p)} \\ \dfrac{px^{p-1}}{\Gamma(p+1)2^p} & \dfrac{2^p(-p)}{x^{p+1}\Gamma(1-p)} \end{vmatrix}\right)$$

$$= \frac{2p}{(\sin \pi p)\Gamma(p+1)\Gamma(1-p)}.$$

A more complete study of the gamma function would show that $\Gamma(p)\Gamma(1-p) = \frac{\pi}{\sin \pi p}$ for *all* p, so $\Gamma(p+1) = p\Gamma(p)$ finally determines

$$W[J_p, Y_p](x) = \frac{2}{\pi x} \neq 0, \tag{3.28}$$

so J_p and Y_p are linearly independent. (In constrast, $W[J_p, J_{-p}](x) = -\frac{2 \sin p\pi}{\pi x}$ is zero whenever p is an integer.) This (finally!) implies that the general solution of Bessel's equation is most conveniently given by the arbitrary linear combination

$$y(x) = k_1 J_p(x) + k_2 Y_p(x) \tag{3.29}$$

for all values p and $x \neq 0$.

Many properties of Bessel functions can be determined by manipulating the series expansions we have obtained. For example, because

$$(J_n(x), x J_n'(x)) = \sum_{k=0}^{\infty} \frac{(-1)^k x^{n+2k}}{k!\Gamma(n+k+1)2^{n+2k}}(1, n+2k),$$

(conveniently writing the two series side by side, as a row vector), shifting indices implies that

$$x J_n'(x) = n J_n(x) + x \sum_{k=1}^{\infty} \frac{(-1)^k x^{n+2k-1}}{(k-1)!\Gamma(n+k+1)2^{n+2k-1}}.$$

Recognizing the latter sum, we obtain the derivative formula

$$x J_n'(x) = n J_n(x) - x J_{n+1}(x). \tag{3.30}$$

Likewise,

$$x J_n'(x) = -n J_n(x) + x J_{n-1}(x), \tag{3.31}$$

so, upon adding and subtracting (3.30) and (3.31), we get the two useful relations

$$J_n'(x) = \frac{1}{2}[J_{n-1}(x) - J_{n+1}(x)] \tag{3.32}$$

and

$$J_n(x) = \frac{x}{2n}[J_{n+1}(x) + J_{n-1}(x)]. \tag{3.33}$$

The *recurrence formula*

$$J_{n+1}(x) = \frac{2n}{x} J_n(x) - J_{n-1}(x) \tag{3.34}$$

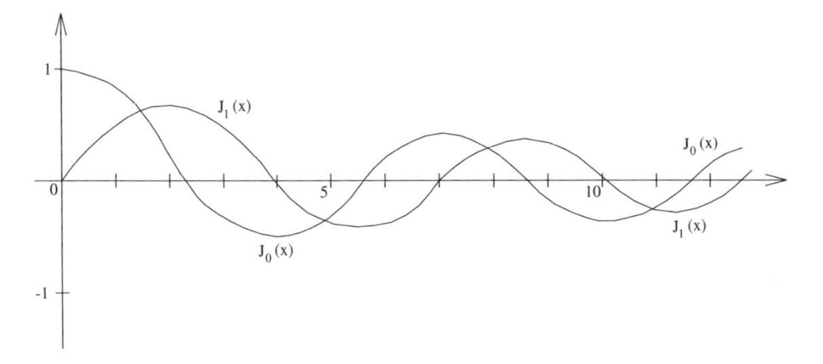

Figure 6: The Bessel functions J_0 and J_1

determines, for example, that

$$J_3(x) = \frac{4}{x} J_2(x) - J_1(x) = \frac{4}{x}\left(\frac{2}{x} J_1(x) - J_0(x)\right) - J_1(x)$$
$$= \left(\frac{8 - x^2}{x^2}\right) J_1(x) - \frac{4}{x} J_0(x).$$

In similar fashion, we can also show that

$$\frac{d}{dx}\left(x^p J_p(x)\right) = x^p J_{p-1}(x). \tag{3.35}$$

Plotting graphs of Bessel functions suggests that they ultimately behave like damped sinusoids. (See Figure 6.). Recall our earlier determination of $J_{\pm\frac{1}{2}}(x)$ and observe that the rescaling $J_m(x) = u(x)/\sqrt{x}$ converts Bessel's equation $J_m'' + \frac{1}{x} J_m' + \left(1 - \frac{m^2}{x^2}\right) J_m = 0$ to

$$u'' + \left(1 + \frac{\frac{1}{4} - m^2}{x^2}\right) u = 0$$

for $x \neq 0$. For large x, then, it is not surprising that we approximately have $J_m(x) \approx \frac{A}{\sqrt{x}} \cos(x + \beta)$, where an amplitude A and a phase shift β must be selected so that the large zeros of J_m and its magnitude nearly agree, providing the approximation

$$J_m(x) \approx \sqrt{\frac{2}{\pi x}} \cos[x - (2m + 1)\pi/4], \tag{3.36}$$

which is actually quite good, even for moderate values of x. The more accurate generalization of this scaling argument as the *WKB* (or *Liouville-Green*) technique applies to many related second-order linear differential equations. For small x, by contrast, a good approximation to J_m results from using the leading term of the series representation, i.e.,

$$J_m(x) \approx \frac{x^m}{2^m \Gamma(m+1)}.$$

Determining the appropriate x ranges to use for various simple approximations is of substantial numerical importance in using such special functions in practice.

Analogous transformations can be used to express other *special functions* in terms of Bessel functions. The *Airy functions* $Ai(x)$ and $Bi(x)$ are linearly independent solutions of the Airy equation

$$y'' + xy = 0. \tag{3.37}$$

Moreover, the simultaneous change of variables

$$y = \sqrt{x}\,u\,(z), \quad z = \frac{2}{3}x^{3/2}$$

converts Airy's equation for $y(x)$ to the Bessel equation

$$z^2 \frac{d^2u}{dz^2} + z\frac{du}{dz} + \left(z^2 - \frac{1}{9}\right) u = 0$$

of order $1/3$ for $u(z)$. Because any $u(z)$ must be a linear combination of $J_{\pm 1/3}$, it follows that any solution of the Airy equation must have the form

$$y(x) = \sqrt{x}\left[\alpha J_{1/3}\left(\frac{2}{3}x^{3/2}\right) + \beta J_{-1/3}\left(\frac{2}{3}x^{3/2}\right)\right] \tag{3.38}$$

for constants α and β. Conjuring up such transformations to Bessel equations is a real art form, but a valuable one since it prevents the unnecessary detailed study and tabulation of even more special functions.

3.4 Legendre polynomials and Sturm-Liouville problems

The *Legendre differential equation*

$$(1 - x^2)y'' - 2xy' + p(p+1)y = 0, \tag{3.39}$$

which often arises in engineering problems with spherical symmetry, becomes singular at $x = \pm 1$. If we seek the Maclaurin expansion

$$y(x) = \sum_{n=0}^{\infty} c_n x^n,$$

for a solution, we would naturally only hope for it to converge when $|x| < 1$. Proceeding as usual, we find that the c_ns must satisfy the recurrence relation

$$c_{n+2} = -\frac{(p-n)(p+n+1)}{(n+2)(n+1)} c_n \quad \text{for each } n \geq 0.$$

Thus, we obtain the general solution

$$\begin{aligned} y(x) = \; & y(0)\left[1 - p(p+1)\frac{x^2}{2!} + (p-2)p(p+1)(p+3)\frac{x^4}{4!} - \ldots\right] \\ & + y'(0)\left[x - (p-1)(p+2)\frac{x^3}{3!}\right. \\ & \left. \qquad + (p-3)(p-1)(p+2)(p+4)\frac{x^5}{5!} + \ldots\right] \end{aligned} \tag{3.40}$$

for $|x| < 1$.

When p is a positive integer, the coefficient c_{p+2} becomes zero, so one of the series in (3.40) (depending on whether p is even or odd) becomes a polynomial of degree p. We can conveniently rewrite it as the (normalized) *Legendre polynomial*

$$P_p(x) = \frac{1}{2^p p!} \frac{d^p}{dx^p}(x^2 - 1)^p, \tag{3.41}$$

if we select c_0 or c_1 so that $c_p = \frac{(2p)!}{2^p (p!)^2}$ and $P_p(1) = 1$. The resulting sequence of polynomials begins with $P_0(x) = 1$, $P_1(x) = x$, $P_2(x) = \frac{1}{2}(3x^2 - 1)$, $P_3(x) = \frac{1}{2}(5x^3 - 3x)$, and $P_4(x) = \frac{1}{8}(35x^4 - 30x^2 + 3)$. Using the differential equation, we could readily show that the polynomials are mutually *orthogonal* in the integrated sense that

$$\int_{-1}^{1} P_j(x) P_k(x) dx = 0 \quad \text{if } j \neq k. \tag{3.42}$$

They are classically used to determine a *best approximation* (in a specified sense) to a given function by a polynomial of fixed degree. Legendre polynomials, however, are only one example of many sets of *orthogonal polynomials* that we often encounter as solutions of *Sturm-Liouville boundary value problems*. They naturally arise in obtaining separated solutions of certain linear partial differential equations in geometrically appropriate regions.

Suppose we, for example, need to determine the temperature $u(x, t)$ in a nonuniform rod with an appropriate heat source [cf. Carslaw and Jaeger (1959)]. We might find that u will satisfy the partial differential equation

$$\rho(x)\frac{\partial u}{\partial t} = \frac{\partial}{\partial x}\left(p(x)\frac{\partial u}{\partial x}\right) + q(x)u \tag{3.43}$$

(on some length interval $a \le x \le b$, for time $t \ge 0$, where ρ, p, and q are given smooth functions with physical significance), together with boundary conditions at endpoints $x = a$ and b and a temperature distribution at $t = 0$. If we seek solutions of (3.43) in the separated form

$$u(x, t) = \phi(x)h(t),$$

(or, by superposition, as their finite, or even infinite, linear combinations), we will need

$$\rho\phi\frac{dh}{dt} = \left[\frac{d}{dx}\left(p\frac{d\phi}{dx}\right) + q\phi\right] h.$$

Then, by separating functions of the two variables x and t, we get

$$\frac{\dot{h}}{h} = \frac{1}{\rho\phi}\frac{d}{dx}\left(p\frac{d\phi}{dx}\right) + \frac{q}{\rho} \equiv -\lambda,$$

which defines the *separation constant* λ (as a simultaneous function of both x and t exclusively). We integrate $\dot{h} = -\lambda h$ to get $h(t) = e^{-\lambda t}h(0)$, which will decay with t whenever $\lambda > 0$. The variable $\phi(x)$ must then satisfy the *Sturm-Liouville eigenvalue problem*, consisting of the ordinary differential equation

$$\frac{d}{dx}\left(p(x)\frac{d\phi}{dx}\right) + q(x)\phi + \lambda\rho(x)\phi = 0 \tag{3.44}$$

on $a < x < b$ together with nontrivial separated boundary conditions, typically of the form

$$\begin{cases} \beta_1\phi(a) + \beta_2\frac{d\phi}{dx}(a) = 0 \\ \text{and} \\ \beta_3\phi(b) + \beta_4\frac{d\phi}{dx}(b) = 0, \end{cases} \tag{3.45}$$

such that $\beta_1^2+\beta_2^2 \neq 0$ and $\beta_3^2+\beta_4^2 \neq 0$ where the βs are obtained from analogous t-independent boundary conditions prescribed for $u(a, t)$ and $u(b, t)$. We will call the resulting *Sturm-Liouville boundary value problem* (3.43, 3.44) *regular* when the coefficients p, q, and ρ are all continuous throughout $[a, b]$, with $p > 0$ and $\rho > 0$ there. We will name the parameter λ an *eigenvalue* if a nontrivial solution ϕ corresponds to it, and will then call the pair $(\lambda, \phi(x))$ an *eigenpair*. Note that any *eigenvector* (or *eigenfunction*) $\phi(x)$ can be multiplied by an arbitrary nonzero constant, which allows it to be normalized in various convenient ways. In addition to these regular problems, we often encounter certain *singular* Sturm-Liouville problems in applications where some of the preceding hypotheses do not hold and some of the significant theory for regular problems is thereby unavailable.

Example 12 The simplest Sturm-Liouville problem may be

$$\begin{cases} u'' + \lambda u = 0, & 0 < x < \pi \\ u(0) = 0, & u(\pi) = 0. \end{cases}$$

The nontrivial real solutions of $u'' + \lambda u = 0$ with $u(0) = 0$ are given by

$$u(x) = c \sin \sqrt{\lambda} x$$

for any $c \neq 0$ and $\lambda > 0$. To have $u(\pi) = 0$ for a nontrivial u, we need $\sin \sqrt{\lambda}\pi = 0$, so $\lambda = n^2$ for $n = 1, 2, \ldots$. The corresponding infinite sequence of eigenpairs

$$(\lambda_n, u_n(x)) \equiv (n^2, \sin nx)$$

provides the basis for *Fourier sine series*, which are of primary importance in many applications, especially in heat conduction and communications engineering (cf. Körner (1988) for a mathematical treatment.).

Example 13 If we next seek bounded solutions of

$$((1 - x^2)u')' + \lambda u = 0$$

on the interval $-1 < x < 1$, we obtain a singular problem because $p(x) \equiv 1 - x^2$ becomes zero at the endpoints and the auxiliary condition (boundedness) is not in the rigidly prescribed form (3.45) of regular boundary conditions. Nonetheless, we readily find that the Legendre polynomials $P_n(x)$ are the only eigenfunctions and that they correspond to the eigenvalues $\lambda_n = n(n + 1)$ for integers $n \geq 0$.

Example 14 Suppose we next seek solutions of

$$\frac{d}{dr}\left(r\frac{du}{dr}\right) + \left(\lambda r - \frac{n^2}{r}\right) u = 0,$$

for a nonnegative integer n and a radial variable $u(r)$ on $0 < r \leq a$, subject to the auxiliary condition that $u(a) = 0$ and that $u(r)$ remains bounded as $r \to 0$. This is again a singular Sturm-Liouville problem for Bessel's equation of order n. Its nontrivial solutions must be multiples of $J_n(\sqrt{\lambda} r)$. Moreover, the boundary condition at $r = a$ requires the value $\sqrt{\lambda_j}\, a$ to be one of the tabulated zeros of J_n.

For a regular Sturm-Liouville problem [(3.44)–(3.45)], standard theory shows that:

(i) There is an infinite sequence of real eigenvalues

$$\lambda_1 < \lambda_2 < \lambda_3 < \ldots,$$

with a smallest eigenvalue λ_1 and with $\lambda_n \to \infty$ as $n \to \infty$.

(ii) The eigenfunction $\phi_n(x)$, corresponding to λ_n, is unique up to an arbitrary constant multiplier, and it has $n - 1$ zeros in the interval $a < x < b$. Moreover, the eigenfunctions $\phi_n(x)$ are orthogonal with respect to the *weight function* $\rho(x)$ in the sense that

$$\int_a^b \rho(x)\phi_n(x)\phi_m(x)dx = 0 \quad \text{for } n \neq m.$$

(iii) The eigenfunctions $\phi_n(x)$ are *complete* in that any piecewise smooth function $f(x)$, with $\int_a^b \rho(x)f^2(x)dx < \infty$, can be "represented" by the *generalized Fourier series*

$$f(x) \sim \sum_{n=1}^{\infty} a_n\phi_n(x)$$

with the coefficients a_n uniquely determined as

$$a_n = \frac{\int_a^b \rho(x)f(x)\phi_n(x)dx}{\int_a^b \rho(x)\phi_n^2(x)dx}.$$

This formula follows directly by integrating $\rho(x)f(x)\phi_n(x)$ termwise, using the orthogonality of the ϕ_js. Convergence of the sum to f holds in the *mean-square* sense, i.e.,

$$\int_a^b \rho(x)[f(x) - f_m(x)]^2dx \to 0 \quad \text{as } m \to \infty$$

for $f_m(x) \equiv \sum_{n=1}^{m} a_n\phi_n(x)$.

We note that much of this theory continues to hold for various singular Sturm-Liouville problems. Accessible presentations of such practical results are contained in Haberman (1987) and Birkhoff and Rota (1978). It comes from an extensive theory, largely developed by European mathematicians early in this century.

3.5 Earthquake protection

The linear boundary value problem consisting of the partial differential equation

$$E^4 \frac{\partial^4 u}{\partial x^2} + \frac{\partial^2 u}{\partial t^2} = \phi, 0 < x < 1, t \geq 0 \tag{3.46}$$

and the boundary conditions

$$\begin{aligned} u(0,t) &= \sin t, u_x(0,t) = 0, \\ u_{xx}(1,t) &= 0, u_{xxx}(1,t) = 0 \end{aligned} \tag{3.47}$$

might describe the periodic vibrations of a building in an earthquake (cf. Fowkes and Mahony (1994) and Segel (1977)). Here, E^4 is a nondimensionalized ratio of the product of Young's modulus and the square of the radius of gyration to the density of the material used.

If we, cleverly, seek a solution in the separated form

$$u(x,t) = V(x)\sin t, \tag{3.48}$$

the transverse displacement V must satisfy the two-point problem

$$\begin{cases} E^4 \dfrac{d^4 V}{dx^4} = V, 0 < x < 1 \\ V(0) = 1, V'(0) = V''(1) = V'''(1) = 0. \end{cases} \tag{3.49}$$

One should think of E as a positive design parameter, seeking to determine how large E needs to be to resist earthquake damage.

For E large, it is natural to simply seek an approximate solution as a regular perturbation series in the small parameter $1/E^4$. Formally setting

$$V\left(x, \frac{1}{E^4}\right) = V_0(x) + \frac{1}{E^4}V_1(x) + \frac{1}{E^8}V_2(x) + \dots,$$

direct substitution into the differential equation and boundary conditions will require that

$$\frac{d^4 V_0}{dx^4} + \frac{1}{E^4}\frac{d^4 V_1}{dx^4} + \dots = \frac{1}{E^4}\left(V_0 + \frac{1}{E^4}V_1 + \dots\right),$$

$$V_0(0) + \frac{1}{E^4}V_1(0) + \dots = 1,$$

$$V_0'(0) + \frac{1}{E^4}V_1'(0) + \dots = 0,$$

$$V_0''(1) + \frac{1}{E^4}V_1''(1) + \dots = 0,$$

$$V_0'''(1) + \frac{1}{E^4}V_1'''(1) + \dots = 0.$$

Equating coefficients successively of powers of $\frac{1}{E^4}$ provides the sequence of boundary value problems

$$\frac{d^4V_0}{dx^4} = 0,\ V_0(0) = 1,\ V_0'(0) = V_0''(1) = V_0'''(1) = 0,$$
$$\frac{d^4V_1}{dx^4} = V_0,\ V_1(0) = V_1'(0) = V_1''(1) = V_1'''(1) = 0,$$

etc.. Solving termwise, we uniquely obtain

$$V_0(x) \equiv 1,\ V_1(x) = \frac{1}{24}x^4 - \frac{1}{6}x^3 + \frac{1}{4}x^2, \text{ etc.}$$

The resulting approximation

$$V(x) = 1 + \frac{x^2}{24E^4}(x^2 - 4x + 6) + \frac{1}{E^8}(\ldots), \tag{3.50}$$

appropriate for large values of E, should be compared numerically to the more-complicated exact solution to be obtained below. It should provide excellent approximations, even when E is only moderately large.

Since the differential equation for V has linearly independent solutions $e^{-x/E}$, $e^{x/E}$, $\sin\frac{x}{E}$ and $\cos\frac{x}{E}$, any solution $V(x)$ must have the form

$$V(x) = \alpha e^{-x/E} + \beta e^{-(1-x)/E} + \gamma \sin\frac{x}{E} + \delta\cos\frac{x}{E} \tag{3.51}$$

where the coefficients α, β, γ, and δ depend on E. Thus, the two initial conditions

$$V(0) = \alpha + \beta e^{-1/E} + \delta = 1$$

and

$$V'(0) = \frac{1}{E}(-\alpha + \beta e^{-1/E} + \gamma) = 0$$

imply that

$$V(x) = \alpha\left[e^{-x/E} + \sin\frac{x}{E} - \cos\frac{x}{E}\right]$$
$$+ \beta\left[e^{-(1-x)/E} - e^{-1/E}\left(\sin\frac{x}{E} - \cos\frac{x}{E}\right)\right] + \cos\frac{x}{E}. \tag{3.52}$$

The terminal conditions thereby provide two remaining equations

$$E^2V''(1) = \alpha\left[e^{-1/E} - \sin\frac{1}{E} + \cos\frac{1}{E}\right]$$
$$+ \beta\left[1 + e^{-\frac{1}{E}}\left(\sin\frac{1}{E} + \cos\frac{1}{E}\right)\right] - \cos\frac{1}{E} = 0$$

and

$$E^3 V'''(1) = \alpha \left[-e^{-1/E} - \cos \frac{1}{E} - \sin \frac{1}{E} \right]$$
$$+ \beta \left[1 + e^{-1/E} \left(\cos \frac{1}{E} - \sin \frac{1}{E} \right) \right] + \sin \frac{1}{E} = 0.$$

Adding and subtracting, we find that α and β must satisfy the linear system

$$-\alpha \sin \frac{1}{E} + \beta \left(1 + e^{-\frac{1}{E}} \cos \frac{1}{E} \right) = \frac{1}{2} \left(\cos \frac{1}{E} - \sin \frac{1}{E} \right)$$
$$\alpha \left(e^{-\frac{1}{E}} + \cos \frac{1}{E} \right) + \beta e^{-\frac{1}{E}} \sin \frac{1}{E} = \frac{1}{2} \left(\cos \frac{1}{E} + \sin \frac{1}{E} \right).$$

The solution will be uniquely determined by Cramer's rule as

$$\alpha = \frac{\cos \frac{1}{E} + \sin \frac{1}{E} - e^{-1/E}}{2\left(\cos \frac{1}{E} \right)(1 + e^{-2/E}) + 4e^{-1/E}}$$

and

$$\beta = \frac{1 + \left(\cos \frac{1}{E} - \sin \frac{1}{E} \right) e^{-1/E}}{2\left(\cos \frac{1}{E} \right)(1 + e^{-2/E}) + 4e^{-1/E}} \tag{3.53}$$

provided E is not one of the infinite sequence of eigenvalues where

$$2e^{-1/E} = -\left(\cos \frac{1}{E} \right)(1 + e^{-2/E})$$

or

$$\cosh \frac{1}{E} = -\sec \frac{1}{E}. \tag{3.54}$$

Near these resonance values, the resulting large amplitudes of α and β predict failure of the structure. Note that graphing both sides of (3.54) shows that the largest eigenvalue occurs for E just below $2/\pi$ and that the eigenvalues become dense as $E \to 0^+$. This prevents the possibility of an asymptotic analysis of solutions for E small, but safety suggests designing with a larger E in any event.

3.6 Exercises

1 Find the Maclaurin series expansion for $\ln(1 + x)$ and its interval of convergence.

2 Determine the Maclaurin series solutions of the following initial value problems and find their sums.

(a) $y' = 1 - y^2$, $y(0) = 0$
(b) $y'' + y = x$, $y(0) = 1$, $y'(0) = 0$

3 Find the first three terms of the Taylor series expansion about $x = 2$ for the solution to

$$y'' + y' - \ln y = e^x, \quad y(2) = e, \quad y'(2) = e^2.$$

4 Determine a procedure to obtain the coefficients in the Taylor series solution of

$$y'' + 3(\sin x)y' - \frac{2x}{1-x}y = 0$$

about any regular point $x_0 (\neq 1)$.

5 Suppose two functions $u(x)$ and $v(x)$ have Taylor series expansions

$$u(x) = a_0 + a_1(x - c) + a_2(x - c)^2 + \dots$$

and

$$v(x) = b_0 + b_1(x - c) + b_2(x - c)^2 + \dots$$

for $|x - c| < r$. Show that

(a) the Wronskian $W[u, v](c)$ of u and v at c is $a_0 b_1 - a_1 b_0$.
(b) u and v are linearly independent on $|x - c| < r$ if $W[u, v](c) \neq 0$.

6 Consider the initial value problem

$$Y' = A(t)Y, \; Y(0) = I$$

where

$$A(t) = \begin{pmatrix} 0 & 1 \\ -1 & -t \end{pmatrix} \text{ and } I = \begin{pmatrix} 1 & 0 \\ 0 & 1 \end{pmatrix}.$$

Find the first three terms of the Maclaurin expansion for the matrix solution $Y(t)$.

7 (a) Show that

$$x^2 y'' + 4xy' + 2y = 0$$

has a general solution of the form $y(x) = \frac{v(x)}{x^2}$ with v smooth for $x > 0$.

(b) Solve the equation

$$x^2y'' + 4xy' + (2 + x^2)y = 0$$

by setting $y(x) = \frac{v(x)}{x^2}$ for $x > 0$.

8 (a) Even though $x = 0$ is a singular point of the equation

$$xy'' - x^2y' + (x^2 - 2)y = 0,$$

find a Maclaurin series solution. Then use reduction of order to determine the form of the general solution.

(b) Repeat (a) for

$$x^2y'' + (x^2 - 3x)y' + 3y = 0.$$

9 (a) The *implicit function theorem* considers algebraic equations

$$f(x, \epsilon) = 0$$

when $f(x, 0) = 0$ has a solution x_0 satisfying $f_x(x_0, 0) \neq 0$. Determine the first few terms of a formal power series solution

$$x(\epsilon) = x_0 + \epsilon x_1 + \epsilon^2 x_2 + \dots,$$

for ϵ small, in terms of the partial derivatives of f, when f, x, and ϵ are all scalars.

(b) Verify the conclusion for

$$x^2 + 2\epsilon x = 1.$$

(c) Explain why the result does not apply to

$$x^2 - \epsilon = 0, \quad \epsilon > 0.$$

10 Find the first two terms of a power series solution $y(x, \epsilon) = \sum_{n=0}^{\infty} a_n(x)\epsilon^n$ of the initial value problem

$$y'' + (1 + \epsilon x)y = 0, \quad y(0) = 1, \quad y'(0) = \epsilon$$

appropriate for small $|\epsilon|$.

11 Solve the equations:

(a) $x^2y'' + 5xy' + 4y = 0, \ x > 0$

(b) $x^2y'' + xy' + y = 0, \ x < 0$

(c) $(3x-1)^2y'' + 3(3x-1)y' - 9y = 0, \ x > 1/3$

12 Find the first three nonzero terms of the Taylor series solution of

$$2x(1+x)y'' + (3+x)y' - xy = 0$$

about the regular singular point $x = -1$.

13 Use the method of Frobenius to find all solutions to

$$x^2y'' + (x^2 - 3x)y' + 3y = 0.$$

(Hint: Because one series solution can be summed, a second can be found by reduction of order.)

14 Suppose the singular first-order scalar equation

$$xy' = a(x)y$$

has a series solution of the form

$$y(x) = x^{a_0} \sum_{k=0}^{\infty} c_k x^k$$

with $c_0 = 1$, for $x > 0$, when $a(x) = \sum_{k=0}^{\infty} a_k x^k$. Find c_1 and c_2.

15 Determine the general solution of the singular system

$$t^2y' = A(t)y$$

of two scalar equations, when

$$A(t) = \begin{pmatrix} -2t & 4 \\ -2t^2 & 5t \end{pmatrix},$$

by using the first equation $t^2y_1' = -2ty_1 + 4y_2$ to eliminate y_2 and the second equation $t^2y_2' = -2t^2y_1 - 5ty_2$ to obtain y_1.

16 Approximate solutions of

$$x^3y'' + \frac{1}{2}x^2y' + y = 0$$

as $x \to \infty$ by letting $z(s) = y(x)$ for $s = 1/x$ near 0.

17 (a) Find two linearly independent solutions of Bessel's equation of order $3/2$ by power series methods.

(b) Find $J_{\pm\frac{3}{2}}(x)$, given that $J_{\frac{1}{2}}(x) = \sqrt{\frac{2}{\pi x}}\sin x$ and $J_{-\frac{1}{2}}(x) = \sqrt{\frac{2}{\pi x}}\cos x$.

18 Multiply the power series for $e^{xt/2}$ and $e^{-x/(2t)}$ to show that

$$e^{\frac{x}{2}(t-\frac{1}{t})} = \sum_{n=-\infty}^{\infty} t^n J_n(x) = J_0(x) + \sum_{n=1}^{\infty} J_n(x)t^n[1+(-1)^n].$$

(This exponential is naturally called the generating function for the Bessel functions J_n.)

19 Use Bessel's equation of order zero to show that J_0' satisfies Bessel's equation of order one. (This, admittedly also follows directly from the identity $2J_0' = J_{-1} - J_1 = -2J_1$.)

20 Determine approximate solutions of

$$y'' + 2\left(\frac{1}{x^3} + \frac{1}{x}\right)y' + \frac{\lambda}{x^4}y = 0$$

for large x by finding series solutions for $z(s) = y(x)$ for small $s = \frac{1}{x}$.

21 Find the first five terms of the Maclaurin series solutions for

(a) $y' = 1 + xy'$, $y(0) = 1$
(b) $y' = \frac{x-y}{x+y}$, $y(0) = 1$
(c) $y' = \frac{1}{y} + 2 + x^2$, $y(0) = 1$

22 Show that one cannot find a Maclaurin series solution of the linear initial value problem

$$y' = y + \sqrt{x},\ y(0) = 1,$$

but if one sets $y(x) = z(u)$ for $u = \sqrt{x}$, there is no problem in finding a Maclaurin series for z as a function of u for $u \geq 0$.

23 (a) To solve the initial value problem

$$\frac{dy}{dx} = \sqrt{y} + \sqrt{x},\ y(0) = 0,$$

introduce $u = \sqrt{y}$ and $t = \sqrt[4]{x}$ to obtain the initial value problem

$$u\frac{du}{dt} + 2t^3(u + t^2),\ u(0) = 0$$

for $u(t)$, when $t \geq 0$.

(b) To find $y = u^2$, seek the Maclaurin expansion for u in the form

$$u(t) = t^3 v(t),$$

where v is the positive solution of

$$3v^2 = 2 + tv\left(2 - \frac{dv}{dt}\right),$$

i.e., $y = \frac{2}{3}t^6\left[1 + tv - \frac{1}{2}tv\frac{dv}{dt}\right]$.

(c) Use successive approximations to "solve" the integral equation

$$y(x) = \int_0^x (\sqrt{y(s)} + \sqrt{s})ds = \frac{2}{3}x^{3/2} + \int_0^x \sqrt{y(s)}\,ds, \quad x \geq 0.$$

Specifically, set $y^{(0)}(x) \equiv 0$ and obtain the first four of the iterates (the superscripts are not derivatives!)

$$y^{(j)}(x) = \frac{2}{3}x^{3/2} + \int_0^x \sqrt{y^{(j-1)}(s)}\,ds, \quad j \geq 1.$$

(d) Set $y(x) = x^{3/2}z(t)$ for $t = x^{1/4}$ and thereby determine the Maclaurin expansion for $z(t)$.

24 Consider the equation

$$xy'' - (x+n)y' + ny = 0.$$

(a) Verify that $y_1(x) = e^x$ is a solution.
(b) Show that y' will satisfy the same equation with n replaced by $n-1$.
(c) When n is a positive integer, verify that

$$y_2(x) = \sum_{j=0}^{n} \frac{x^j}{j!}$$

is a second linearly independent solution of the differential equation.
(d) Note that the Wronskian $W[y_1, y_2](0) = 0$. Doesn't this contradict the linear independence of y_1 and y_2?

25 Obtain the first five coefficients in the Taylor series about $x = 1$ for the solution to the initial value problem

$$x^2y''' + xy' - y = 0, \ y(1) = 1, \ y'(1) = 0, \ y''(1) = 1$$

and determine its interval of convergence.

26 (a) Show that the Maclaurin series for the solution to

$$(1+x^2)y'' - 3xy' + 3y = 0$$

converges for $|x| < 1$ (being limited by the singular points at $x = \pm i$).

(b) Determine the radius of convergence of the solution given by the Taylor series expansion about $x = 1$.

27 Determine the first four terms of the Maclaurin series for the solution of the coupled nonlinear system

$$\begin{cases} \dfrac{dx}{dt} = ty, & x(0) = 1 \\ \dfrac{dy}{dt} = xy, & y(0) = 1. \end{cases}$$

28 Find series solutions of $xy'' - y = 0$.

29 Show that the modified Bessel equation

$$t^2\ddot{x} + t\dot{x} - (t^2 + p^2)x = 0$$

of order $p \geq 0$ has a solution

$$I_p(t) = \sum_{n=0}^{\infty} \frac{1}{n!\Gamma(n+p+1)} \left(\frac{t}{2}\right)^{2n+p}.$$

30 (a) Show that the differential equation

$$\ddot{x} + e^{-t}x = 0$$

(which models an *aging spring*) can be converted to Bessel's equation of order zero by introducing the new independent variable $s = e^{-t}$.

(b) Use the limiting differential equation (as $t \to \infty$) to predict the behavior of solutions $x(t)$ as $t \to \infty$.

(c) Presuming $J_0(0)$ is nonzero but bounded, show why some solutions of Bessel's equation of order zero must be singular (like $\log |z|$) as $z \to 0$.

31 Obtain series solutions to

(a) $x^2y'' + (\sin x)y' + (\cos x)y = 0$

(b) $x^2y'' + xy' - y = x$

32 Determine the Maclaurin series for the smooth solution of the initial value problem

$$y'' + \left(x + \frac{2}{x}\right) y' + 2y = 0, \quad y(0) = 1$$

and show that this solution is given by *Dawson's integral*

$$y(x) = \frac{1}{x} \int_0^x e^{\frac{1}{2}(s^2 - x^2)} ds.$$

33 Find the general solution to

$$y'' + \frac{y'}{(x-a)^2} = 0$$

for $x \neq a$.

34 Use the transformations $u = x^a y$ and $z = bx^c$ to show that a solution of

$$x^2 u'' + (1 - 2a)xu' + (b^2c^2x^{2c} + a^2 - k^2c^2)u = 0$$

is given by $u(x) = x^a J_k(bx^c)$. Show how this result determines the solution of Airy's equation in terms of Bessel functions of order $\pm 1/3$.

35 For $x^2y'' + xy' + (1+x)y = 0$, find two linearly independent real solutions in the form

$$y(x) = \left(\sum_{n=0}^{\infty} a_n x^n\right) \sin(\ln x) + \left(\sum_{n=0}^{\infty} b_n x^n\right) \cos(\ln x), \ x > 0.$$

36 By differentiating under the integral sign, show that the integral

$$y(x) = \frac{1}{\pi} \int_0^{\pi} \cos(x \sin\theta) d\theta$$

satisfies $y'(x) = -\frac{x}{\pi} \int_0^{\pi} \cos(x \sin\theta) \cos^2\theta d\theta$ for $x \geq 0$. Further show that y satisfies Bessel's equation of order zero. Then, use the uniqueness theorem to show that $y(x) = J_0(x)$.

37 Suppose $y(t)$ satisfies Bessel's equation $t^2\ddot{y} + t\dot{y} + (t^2 - p^2)y = 0$. Show that

$$z(x) = e^{ax} y(bx)$$

for real constants a and b, with $b \neq 0$, satisfies

$$x^2 z'' + x(1 - 2ax)z' + [(a^2 + b^2)x^2 - ax - p^2]z = 0.$$

38 (a) Solve $x^2y'' + ky = 0$ on $x > 0$.
(b) Show that solutions have a finite number of zeros if the constant $k \leq 1/4$,
(c) Show that solutions have an infinite number of zeros if $k > 1/4$.

39 (a) Use the method of Frobenius to solve

$$xy'' + (x-2)y' - 3y = 0.$$

(b) Determine a closed-form expression for the general solution for $x \neq 0$.

40 By guessing that a particular solution is a polynomial, solve the singular equations

$$x^2y'' - 6y = 2x^2$$

and

$$x^2y'' - 4xy' + 6y = 2x + 5$$

(Recall that another solution method is to introduce $t = \ln x$).

41 (a) Use the method of Frobenius to solve

$$4xy'' + 2y' - y = 0$$

in terms of hyperbolic functions.

(b) Relate your answer to the solutions $y = e^{\pm\sqrt{x}}$.

42 Show that

$$y = \sqrt{x}\,J_2(4\sqrt[4]{x})$$

satisfies

$$x^{3/2}y'' + y = 0 \quad \text{for} \quad x > 0.$$

43 Consider the scalar initial value problem

$$\dot{x} = a(t)x,$$

where $a(t)$ has a Maclaurin expansion $a(t) = \sum_{k=0}^{\infty} a_k t^k$, and $x(0) = 1$.

(a) Determine the recursion relation for the terms of the Maclaurin expansion

$$x(t) = \sum_{k=0}^{\infty} c_k t^k$$

of the solution.

(b) Find the exact solution and show that the first four coefficients of the series obtained agree with those found through the Maclaurin expansion of the exact solution.

(c) Suppose the scalar functions a, b, c, f, and g in the system below each have given Maclaurin series expansions. Determine recursion relations for the terms of the Maclaurin expansions

$$(x(t), y(t)) = \sum_{k=0}^{\infty} (c_k, d_k)t^k$$

for the solution to the coupled initial value problem

$$\begin{cases} \dot{x} = a(t)x + f(t), \ x(0) = 1 \\ \dot{y} = b(t)x + c(t)y + g(t), \ y(0) = 0. \end{cases}$$

44 Consider the vector-matrix system

$$x' = A(t)x + b(t),$$

where $A(t)$ and $b(t)$ have convergent Maclaurin expansions. Find the recurrence relation for successive coefficients in the Maclaurin expansion

$$x(t) = \sum_{k=0}^{\infty} c_k t^k$$

of the solutions. Note that $x(0) = c_0$ is arbitrary.

45 (a) Consider the linear matrix initial value problem

$$X' = (A_0 + A_1 t)X, \ X(0) = I$$

and determine successive coefficients in the power series solution

$$X(t) = \sum_{k=0}^{\infty} c_k t^k \quad \text{with } c_0 = I.$$

(b) Sum the series obtained when the matrices A_0 and A_1 commute, i.e., $A_0 A_1 = A_1 A_0$.

46 Find solutions of

$$x^4 y'' + 2x^3 y' - y = 0$$

of the form $y = \sum_{n=0}^{\infty} \frac{a_n}{x^n}$ for large $x > 0$.

47 The Thomas-Fermi theory for ionized gases uses the solution of the two-point boundary value problem

$$y'' - \sqrt{\frac{y^3}{x}} = 0, \ \ y(0) = 1, \ \ y(1) = 0.$$

Seek a "formal" series solution

$$y(x) = 1 + \sum_{n=0}^{\infty} a_n x^{r+np}$$

for $a_0 \neq 0$ and $p \neq 0$.

48 *Euler's method* to numerically approximate the solution of the initial value problem

$$y' = f(x, y), \quad y(x_0) = y_0$$

consists of defining $x_{n+1} = x_n + h$ for some small $h > 0$ and of approximating $y_{n+1} = y(x_{n+1})$ by using the first two terms of the Maclaurin expansion for $y(x_n + h)$ with $y'(x_n) = f(x_n, y_n)$. Approximate the solution of the linear example $y' = 1 - x + 4y$, $y(0) = 1$ on $0 \le x \le 1$ using $h = 0.1$ and $n = 0, 1, \ldots, 10$. What error is made at $x = 1/2$ and $x = 1$?

49 (a) Show that the product

$$u(r, t, \omega) = \cos(\omega t) J_0(\omega r)$$

satisfies the partial differential equation

$$\frac{\partial^2 u}{\partial r^2} + \frac{1}{r}\frac{\partial u}{\partial r} = \frac{\partial^2 u}{\partial t^2}$$

for any constant ω.

(b) Show that the infinite series

$$u(r, t) \equiv \sum_{k=0}^{\infty} c_k u(r, t; \omega_k)$$

will "formally" satisfy the same partial differential equation as well as the homogeneous boundary condition $u(1, t) = 0$, provided the *eigenvalues* ω_k are selected to be roots of the transcendental equation

$$J_0(\omega) = 0$$

(tabulated, for example, in Abramowitz and Stegen (1965)).

(c) Suppose the initial function $u(r, 0)$ can be expanded as $u(r, 0) = \sum_{k=0}^{\infty} c_k u(r, 0; \omega_k)$ for some constants c_k. Show that the resulting series $u(r, t)$ then satisfies the differential equation, the homogeneous boundary condition at $r = 1$, and the initial condition at $t = 0$.

50 The van der Pol oscillator (which arises in both electronic and biological applications) obeys the differential equation

$$u'' + \epsilon(u^2 - 1)u' + u = 0.$$

For ϵ small, one might anticipate having solutions with a period of nearly 2π in t. Thus, introduce the new time

$$T = t/\omega$$

and seek a periodic solution of the form

$$u(T,\epsilon) = u_0 + \epsilon u_1 + \epsilon^2 u_2 + \dots$$

with

$$\omega = 1 + \epsilon\omega_1 + \epsilon^2\omega_2 + \dots,$$

and show that one gets a unique periodic solution such that

$$u = 2\cos\omega T + \frac{\epsilon}{4}\sin 3\omega T + \epsilon^2(\cdots)$$

and

$$\omega = 1 - \frac{\epsilon^2}{16} + \epsilon^3(\cdots).$$

51 By introducing $s = 1/\sqrt{x}$, show that Legendre's equation

$$(1-x^2)y'' - 2xy' + n(n+1)y = 0$$

has a regular singular point at $x = \infty$ $(s = 0)$.

52 Find vector solutions of the linear system

$$x' = \left(A_0 + \frac{1}{t}A_1 + \frac{1}{t^2}A_2\right)x$$

for large t as the series

$$x(t) = e^{\lambda t}\sum_{k=0}^{\infty}\frac{C_k}{t^{k+r}}$$

when

$$A_0 = \begin{pmatrix} 0 & 1 \\ 0 & -1 \end{pmatrix},\ A_1 = \begin{pmatrix} 0 & 0 \\ 0 & 1 \end{pmatrix},\ \text{and } A_2 = \begin{pmatrix} 0 & 0 \\ -1 & 0 \end{pmatrix}.$$

53 Consider the differential equation

$$y'' + p(x)y' + q(x)y = 0$$

where p and q have the expansions

$$(p(x), q(x)) = (p_0, q_0) + \frac{1}{x}(p_1, q_1) + \dots$$

at infinity. Show, formally, that y has a solution

$$y(x) = e^{\lambda x}x^{\delta}v(x)$$

where $v(x) = v_0 + \frac{1}{x}v_1 + \dots$ whenever $\lambda^2 + p_0\lambda + q_0 = 0$ and $\delta(2\lambda + p_0) + \lambda p_1 + q_1 = 0$.

54 Given that $e^{1/x}$ satisfies

$$x^3y'' + xy' - 2y = 0,$$

explain why no solution will have a power series expansion about $x = 0$.

Solutions to exercises 1–20

1 If $g(x) = \ln(1 + x)$, then $g(0) = 0$, $g'(0) = \left(\frac{1}{1+x}\right)\Big|_{x=0} = 1$, $g''(0) = -\frac{1}{(1+x)^2}\Big|_{x=0} = -1$, $g'''(0) = \frac{2}{(1+x)^3}\Big|_{x=0} = 2$, etc. Thus, $g(x)$ has the Maclaurin expansion $g(x) = x - \frac{1}{2}x^2 + \frac{1}{3}x^3 - \frac{1}{4}x^4 + \ldots$. The ratio test provides $R = \lim_{n\to\infty}\left|\frac{\frac{1}{n+1}}{\frac{1}{n}}\right| = 1$, reflecting the singularity of g at $x = -1$.

2 (a) The equation $y' = 1 - y^2$ is separable. Because $\frac{dy}{1-y^2} = dx$, $x + c = \tan^{-1} y$ for some c. Taking $y(0) = 0$, we get $y(x) = \tan x$ for $|x| < \pi/2$. Note that $y(0) = 0$ implies that $y'(0) = 1$. Differentiating the equation twice yields $y''(0) = 0$ and $y'''(0) = 2$, so $y(x) = x - \frac{1}{3}x^3 + \ldots$. More systematically, setting $y = \sum_{n=0}^{\infty} a_n x^n$ implies that $y' = \sum_{n=1}^{\infty} na_n x^{n-1} = \sum_{m=0}^{\infty}(m + 1)a_{m+1}x^m$ and $y^2 = \sum_{n=0}^{\infty} x^n \sum_{j=0}^{n} a_j a_{n-j}$ (by the Cauchy product formula). Now $a_0 = 0$, $a_1 = 1$, $a_2 = 0$, and $a_3 = -1/3$. Equating coefficients successively requires $(m + 1)a_{m+1} = \sum_{j=0}^{m} a_j a_{m-j}$ for $m \geq 2$. Moreover, $a_0 = 0$ implies that $a_{m+1} = \left(\sum_{j=1}^{m-1} a_j a_{m-j}\right)/(m + 1)$, i.e., $a_3 = \frac{a_1^2}{3} = \frac{1}{3}$, $a_4 = \frac{a_1 a_2}{2} = 0$, $a_5 = \frac{1}{5}(2a_1a_3 + a_2^2) = \frac{2}{15}$, etc. Convergence must be limited to $|x| < \frac{\pi}{2}$ since $\cos\frac{\pi}{2} = 0$ causes $\tan\left(\pm\frac{\pi}{2}\right)$ to be undefined.

(b) By undetermined coefficients, the unique solution of $y'' + y = x$, $y(0) = 1$, $y'(0) = 0$ is given by $y(x) = x + \cos x - \sin x$. If we seek a series solution $y(x) = \sum_{n=0}^{\infty} a_n x^n$, then $y''(x) = \sum_{n=2}^{\infty} n(n - 1)a_n x^{n-2} = \sum_{m=0}^{\infty}(m + 2)(m + 1)a_{m+2}x^m$ implies that we need $\sum_{j=0}^{\infty}[(j + 2)(j + 1)a_{j+2} + a_j]x^j = x$, while the initial conditions imply that $a_0 = y(0) = 1$ and $a_1 = y'(0) = 0$. Equating coefficients successively, then, we need $2a_2 + a_0 = 0$, $6a_3 + a_1 = 1$, and $(j + 2)(j + 1)a_{j+2} + a_j = 0$ for each $j \geq 2$. Thus, $a_2 = -\frac{1}{2}$, $a_3 = \frac{1}{6}$, $a_4 = \frac{1}{24}$, etc., and $y(x) = 1 - \frac{1}{2}x^2 + \frac{1}{6}x^3 + \frac{1}{24}x^4 + \ldots = x + \cos x - \sin x$ for all x.

3 The initial value problem $y'' + y' - \ln y = e^x$, $y(2) = e$, $y'(2) = e^2$ has a solution with the Taylor series $y(x) = y(2) + y'(2)(x - 2) + \frac{1}{2}y''(2)(x - 2)^2 + \ldots$ near $x = 2$. Since we are given $y(2)$ and $y'(2)$, the differential

equation provides $y''(2) = e^2 + \ln(y(2)) - y'(2) = 1$, and the series begins like $e + e^2(x-2) + \frac{1}{2}(x-2)^2 + \ldots$. Further coefficients can be obtained by differentiating the equation repeatedly and evaluating the result at $x = 2$.

4 Note the Taylor series $\sin x = \sin x_0 + \cos x_0(x - x_0) - \frac{1}{2}\sin x_0(x-x_0)^2 + \ldots \equiv \sum_{j=0}^{\infty} b_j(x - x_0)^j$, which converges for all x, while

$$-\frac{x}{1-x} = 1 - \frac{1}{1-x} = 1 - \frac{1}{1-x_0}\left(1 - \left(\frac{x-x_0}{1-x_0}\right)\right)^{-1} =$$

$$1 - \frac{1}{1-x_0}\left(1 + \frac{x-x_0}{1-x_0} + \frac{(x-x_0)^2}{(1-x_0)^2} + \ldots\right) \equiv \sum_{j=0}^{\infty} c_j(x-x_0)^j$$

for $|x - x_0| < |1 - x_0|$, where $c_0 = -\frac{x_0}{1-x_0}$ and $c_j = -(1-x_0)^{-(j+1)}$ for $j \geq 1$. (Note: It would be easier to get the b_js by writing $\sin x = \sin(x_0 + x - x_0) = \sin x_0 \cos(x - x_0) + \cos x_0 \sin(x - x_0)$ and then using the Taylor series for $\cos(x - x_0)$ and $\sin(x - x_0)$.) If we set $y(x) = \sum_{k=0}^{\infty} a_k(x - x_0)^k$ for $|x - x_0| < |1 - x_0|$, substitution into the differential equation will require

$$\sum_{k=0}^{\infty}(k+2)(k+1)a_{k+2}(x-x_0)^k$$
$$+ 3\left[\sum_{k=0}^{\infty} b_k(x-x_0)^k\right]\left[\sum_{k=0}^{\infty}(k+1)a_{k+1}(x-x_0)^k\right]$$
$$+ 2\left[\sum_{k=0}^{\infty} c_k(x-x_0)^k\right]\left[\sum_{k=0}^{\infty} a_k(x-x_0)^k\right] = 0.$$

Equating coefficients of $(x - x_0)^k$ for each $k \geq 0$ therefore requires

$$(k+2)(k+1)a_{k+2} + 3\sum_{j=0}^{k} b_{k-j}(j+1)a_{j+1} + 2\sum_{j=0}^{k} c_{k-j}a_j = 0.$$

Thus, each a_{k+2} becomes successively determined as a function of the unspecified $a_1 = y'(x_0)$ and $a_0 = y(x_0)$. Thus, $y(x) = y(x_0)[1 + \frac{x_0}{1-x_0}(x - x_0)^2 + \ldots] + y'(x_0)[(x - x_0) - \frac{3}{4}\sin x_0(x - x_0)^3 + \ldots]$.

5 (a) Given the Taylor series $u(x) = a_0 + a_1(x - c) + a_2(x - c)^2 + \ldots$ and $v(x) = b_0 + b_1(x - c) + b_2(x - c)^2 + \ldots$, their Wronskian at c is given by

$$W(u, v)(c) = \begin{vmatrix} u(c) & v(c) \\ u'(c) & v'(c) \end{vmatrix} = \begin{vmatrix} a_0 & b_0 \\ a_1 & b_1 \end{vmatrix} = a_0 b_1 - a_1 b_0.$$

(b) Consider the linear combination $\alpha u(x) + \beta v(x) = 0$. Equating coefficients of all powers $(x-c)^j$ requires that $\alpha a_j + \beta b_j = 0$ for every j. The first two conditions $\alpha a_0 + \beta b_0 = 0$ and $\alpha a_1 + \beta b_1 = 0$ form a linear system for α and β, which has the unique solution $\alpha = \beta = 0$ since its determinant of coefficients is $a_0 b_1 - a_1 b_0 \neq 0$. This implies the linear independence of the given functions u and v.

6 Any $Y(t)$ will have the Maclaurin series $Y(t) = Y(0)+Y'(0)t+\frac{1}{2}Y''(0)t^2+ \dots$, at least near $t = 0$. If Y is defined as the 2×2 matrix solution of the initial value problem $Y' = A(t)Y$, $Y(0) = I$ for $A(t) = A(0) + A'(0)t$ with

$$A(0) + A'(0)t = \begin{pmatrix} 0 & 1+t \\ -1 & -t \end{pmatrix},$$

equating coefficients of successive powers of t implies that $Y(0) = I$, $Y'(0) = A(0)$, and

$$Y''(0) = A(0)Y'(0) + A'(0)Y(0) = A^2(0) + A'(0) = \begin{pmatrix} -1 & 0 \\ 0 & -2 \end{pmatrix}.$$

Thus,

$$Y(t) = \begin{pmatrix} 1 & 0 \\ 0 & 1 \end{pmatrix} + \begin{pmatrix} 0 & 1 \\ -1 & 0 \end{pmatrix} t + \frac{1}{2}\begin{pmatrix} -1 & 0 \\ 0 & -2 \end{pmatrix} t^2 + \dots, \quad \text{so,}$$

$$Y(t) = \begin{pmatrix} 1 - \frac{t^2}{2} + \dots & t + \dots \\ -t + \dots & 1 - t^2 + \dots \end{pmatrix}.$$

Those seeking a further challenge could explicitly solve the problem by examining the coupled system for the four components of $Y(t)$ and reducing it to explicitly solving the resulting pair of coupled second-order equations.

7 (a) The Cauchy-Euler equation $x^2y'' + 4xy' + 2y = 0$ has solutions of the form x^r, where $r(r-1)+4r+2 = 0$. Thus, the general solution $y(x) = \frac{c_1}{x^2} + \frac{c_2}{x}$ for $x \neq 0$ has the form $y = \frac{v(x)}{x^2}$, where $v = c_1 + c_2 x$ satisfies $v'' = 0$.

(b) If we seek a solution of $x^2y'' + 4xy' + (2+x^2)y = 0$ of the form $y = \frac{v(x)}{x^2}$, v must satisfy $v'' + v = 0$, so the general solution for v shows that $y(x) = \frac{1}{x^2}(c_1 \cos x + c_2 \sin x)$ for $x \neq 0$.

8 (a) If we seek a solution of the singular equation $xy'' - x^2y' + (x^2-2)y = 0$ as the Maclaurin series $y = \sum_{n=0}^{\infty} a_n x^n$, formal substitution into the equation requires that $-2a_0 + (-2a_1 + 2a_2)x + \sum_{m=2}^{\infty}[(m+2)ma_{m+1} - (m-1)a_{m-1} + a_{m-2} - 2a_m]x^m = 0$. Equating coefficients successively implies that $a_0 = 0$, $a_2 = a_1$, $6a_3 = 2a_2 + a_1$, $12a_4 =$

$2a_3 + 2a_2 - a_1$, etc. Thus, we get the series solution $y(x) = a_1(x + x^2 + \frac{1}{2}x^3 + \frac{x^4}{6} + \ldots)$, which we identify as the sum $a_1 xe^x$. That xe^x is a solution can be directly verified. Using reduction of order, we next show that any solution $y_2(x) = xe^x v(x)$ requires v to satisfy $xv'' = (x^2 - 2x - 2)v'$, so, upon integration, $v(x) = \int^x \frac{1}{s^2} e^{\frac{s^2}{2} - 2s} ds$. The expansion of y_2 about $x = 0$ can be obtained by formally integrating (the Laurent expansion of) the integrand about $s = 0$. Thus, $y_2(x) = xe^x \int^x \frac{1}{s^2} e^{s^2/2} e^{-2s} ds = -2x \ln|x| + (-1 - x + 2x^2 + \ldots)$, in the form the method of Frobenius predicts.

(b) If we substitute the series $y(x) = \sum_{n=0}^{\infty} a_n x^n$ into the differential equation $x^2 y'' + (x^2 - 3x)y' + 3y = 0$, we get $3a_0 + 0x - a_2 x^2 + \sum_{m=3}^{\infty} x^m (m-1)[(m-3)a_m + a_{m-1}] = 0$. Thus, a_0 and a_2 are zero, and $(m-3)a_m = a_{m-1}$ for each $m \geq 3$. Taking $a_1 = 0$ and leaving a_3 arbitrary therefore provides the series $y_1(x) = a_3 x^3 (1 - x + \frac{1}{2}x^2 - \ldots)$, which sums to $a_3 x^3 e^{-x}$. We can find another linearly independent solution $y_2(x) = x^3 e^{-x} v(x)$, provided $x^5 v'' + (3x^4 - x^5)v' = 0$. Thus, we get $y_2(x) = x^3 e^{-x} \int^x \frac{e^s}{s^3} ds$. Its expansion also involves the logarithm predicted by Frobenius.

9 (a) Define $F(\epsilon) \equiv f(x(\epsilon), \epsilon) = 0$ for $x(\epsilon) = x_0 + \epsilon x_1 + \epsilon^2 x_2 + \ldots$. Using the Maclaurin series $F(\epsilon) = F(0) + F'(0)\epsilon + \frac{1}{2}F''(0)\epsilon^2 + \ldots$ in ϵ requires $F^j(0) = 0$ for all j. First, note that $F(0) = f(x_0, 0) = 0$ is locally uniquely solvable for x_0, provided $f_x(x_0, 0) \neq 0$. Next,

$$F'(\epsilon) = f_x(x(\epsilon), \epsilon) \frac{dx_\epsilon}{d\epsilon} + f_\epsilon(x(\epsilon), \epsilon)$$

for $\epsilon = 0$ implies that

$$f_x(x_0, 0)x_1 + f_\epsilon(x_0, 0) = 0,$$

so $x_1 = -f_x^{-1}(x_0, 0) f_\epsilon(x_0, 0)$. Continuing,

$$\begin{aligned} F''(\epsilon) = \; & f_{xx}(x(\epsilon), \epsilon) \left(\frac{dx_\epsilon}{d\epsilon} \right)^2 + 2 f_{x\epsilon}(x(\epsilon), \epsilon) \frac{dx_\epsilon}{d\epsilon} \\ & + f_{\epsilon\epsilon}(x(\epsilon), \epsilon) + f_x(x(\epsilon), \epsilon) \frac{d^2 x_\epsilon}{d\epsilon^2} \end{aligned}$$

implies for $\epsilon = 0$ that

$$f_{xx}(x_0, 0)x_1^2 + 2 f_{x\epsilon}(x_0, 0)x_1 + f_{\epsilon\epsilon}(x_0, 0) + 2 f_x(x_0, 0)x_2 = 0,$$

so we solve for x_2 directly in terms of x_0 and x_1, which we already know in terms of x_0. Corresponding to any isolated root x_0

of $f(x_0, 0) = 0$, we therefore have its unique regular perturbation series

$$x(\epsilon) = x_0 - \epsilon f_x^{-1}(x_0, 0) f_\epsilon(x_0, 0) - \epsilon^2 f_x^{-1}(x_0, 0)[f_{xx}(x_0, 0)$$
$$(f_x^{-1}(x_0, 0) f_\epsilon(x_0, 0))^2 + 2 f_{x\epsilon}(x_0, 0) f_x^{-1}(x_0, 0) f_\epsilon(x_0, 0)$$
$$+ f_{\epsilon\epsilon}(x_0, 0)] + \epsilon^3(\ldots).$$

Further terms x_j can be found successively. To go much further, however, use of Maple or Mathematica is recommended.

(b) The quadratic $x^2 + 2\epsilon x = 1$ has the two solutions $x(\epsilon) = -\epsilon \pm \sqrt{1+\epsilon^2}$. For $|\epsilon| < 1$, the binomial theorem provides the convergent expansions $x(\epsilon) = \pm 1 - \epsilon \mp \frac{1}{2}\epsilon^2 + \ldots$. These can be as easily obtained by formally substituting the power series into the given polynomial and equating coefficients of like powers of ϵ. Thus, $(x_0 + \epsilon x_1 + \epsilon^2 x_2 + \ldots)^2 + 2\epsilon(x_0 + \epsilon x_1 + \epsilon^2 x_2 + \ldots) = x_0^2 + 2\epsilon x_0(x_1 + 1) + \epsilon^2(2x_0x_2 + x_1^2 + 2x_2) + \ldots = 1$ implies that $x_0^2 = 1$, $x_1 = -1$, $x_2 = -\frac{x_1^2}{2x_0+1}$, etc. Two solutions are obtained, without complications, since $f_x(x_0, 0) = 2x_0 \neq 0$ for both roots $x_0 = 1$ and -1 of $x_0^2 = 1$.

(c) The equation $x^2 - \epsilon = 0$ has the two real solutions $x = \pm\sqrt{\epsilon}$, provided $\epsilon > 0$. Their expansions are in powers of $\sqrt{\epsilon}$, rather than ϵ, reflecting the breakdown occurring because $x_0^2 = 0$ has only the trivial solution on which $f_x(x_0, 0)$ is zero. (Solutions to arbitrary polynomials with ϵ-dependent coefficients can be found in terms of fractional powers of ϵ using the so-called *Newton polygon method.*)

10 Setting $y(x, \epsilon) = a_0(x) + \epsilon a_1(x) + \ldots$ and differentiating termwise, we formally obtain $(a_0'' + \epsilon a_1'' + \ldots) + (1 + \epsilon x)(a_0 + \epsilon a_1 + \ldots) = 0$, $a_0(0) + \epsilon a_1(0) + \ldots = 1$, and $a_0'(0) + \epsilon a_1'(0) + \ldots = \epsilon$. Equating coefficients of ϵ^0 and ϵ^1 thereby provides the initial value problems $a_0'' + a_0 = 0$, $a_0(0) = 1$, $a_0'(0) = 0$ and $a_1'' + a_1 + xa_0 = 0$, $a_1(0) = 0$, $a_1'(0) = 1$. Thus, $a_0(x) = \cos x$, $a_1(x) = \frac{1}{4}(-x\cos x + (5 - x^2)\sin x)$, etc. We can expect the truncated series to provide good approximate solutions for $|\epsilon|$ sufficiently small on bounded x intervals.

11 (a) If we seek a solution to $x^2y'' + 5xy' + 4y = 0$ as $y(x) = x^r$, r must satisfy $r(r-1) + 5r + 4 = (r+2)^2 = 0$. Reduction of order then provides the general solution $y(x) = \frac{1}{x^2}(c_1 + c_2 \ln x)$ for $x > 0$.

(b) To solve $x^2y'' + xy' + y = 0$ for $x < 0$, we might set $y(-x) = z(s)$ for $s > 0$ and seek solutions to $s^2\frac{d^2z}{ds^2} + s\frac{dz}{ds} + z = 0$ of the form $z(s) = s^r$, where $r(r-1)+r+1 = r^2+1 = 0$. Taking $z(s) = s^{\pm i} = e^{\pm i \ln s} = \cos(\ln s) \pm i \sin(\ln s)$, we get the general real solution $y(x) = c_1 \cos(\ln(-x)) + c_2 \sin(\ln(-x))$ for $x < 0$.

(c) To solve $(3x-1)^2 y'' + 3(3x-1)y' - 9y = 0$ for $x > 1/3$, we set $y = (3x-1)^r$. Then, $(3x-1)y' = 3r(3x-1)^r$ and $(3x-1)^2 y'' = 9r(r-1)(3x-1)^r$, so $9r(r-1)+3(3r)-9 = 9(r^2-1) = 0$. Since $r = 1$ or -1, we get the general solution $y(x) = c_1(3x-1) + \frac{c_2}{3x-1}$, provided $x \neq 1/3$.

12 To study solutions of $2x(1+x)y'' + (3+x)y' - xy = 0$ near the regular singular point $x = -1$, we introduce $z = x+1$ and seek a solution of the transformed equation $(-2z+2z^2)\frac{d^2y}{dz^2} + (2+z)\frac{dy}{dz} + (1-z)y = 0$ as the Maclaurin series $y(z) = \sum_{n=0}^{\infty} c_n z^n$. Substituting, we obtain $(c_0 + 2c_1) + (2c_1 - c_0)z + \sum_{m=2}^{\infty}[-2(m^2-1)c_{m+1} + (2m^2 - m + 1)c_m - c_{m-1}]z^m = 0$. Then, $c_0 + 2c_1 = 2c_1 - c_0 = 0$ implies that $c_0 = c_1 = 0$, while $2(m^2-1)c_{m+1} = (2m^2 - m + 1)c_m - c_{m-1}$ for $m \geq 2$ provides the solution $y(x) = c_2(1+x)^2[1 + \frac{7}{6}(1+x) + \frac{53}{48}(1+x)^2 + \ldots]$, with c_2 arbitrary. The linearly independent solutions are singular at $x = -1$.

13 If we seek a solution of $x^2 y'' + x(-3+x)y' + 3y = 0$ in the form $y(x) = \sum_{n=0}^{\infty} c_n x^{n+r}$ with $c_0 \neq 0$, r will satisfy the indicial equation $r(r-1) - 3r + 3 = (r-3)(r-1) = 0$, so we take $y_1(x) = \sum_{n=0}^{\infty} c_n x^{n+3}$, where $\sum_{n=0}^{\infty}(n+2)(nc_n + c_{n-1})x^{n+3} = 0$ provides $c_n = \frac{(-1)^n}{n!}c_0$ for $n \geq 0$. Thus, $y_1(x) = \sum_{n=0}^{\infty} \frac{(-1)^n}{n!} x^{n+3} = x^3 e^{-x}$ is a solution, and we can obtain the general solution in the form $y(x) = x^3 e^{-x} u(x)$, provided u satisfies $u'' + (\frac{3}{x} - 1)u' = 0$. Solving for u, we obtain $y(x) = x^3 e^{-x} c_1 + x^3(\int^x \frac{e^{-x+s}}{s^3})c_2$. Expanding y about $x = 0$, we find a logarithmic singularity there unless $c_2 = 0$.

14 If we seek a solution of $xy' = a(x)y$, with $a(x)$ having the Maclaurin series $\sum_{k=0}^{\infty} a_k x^k$, in the form $y = x^{a_0}(1 + \sum_{k=1}^{\infty} c_k x^k)$, we will need $xy' = a_0 x^{a_0} + c_1(a_0+1)x^{a_0+1} + c_2(a_0+2)x^{a_0+2} + \ldots = a(x)y = a_0 x^{a_0} + (a_1 + a_0 c_1)x^{a_0+1} + (a_2 + a_1 c_1 + a_0 c_2)x^{a_0+2} + \ldots$. Equating coefficients requires $c_1 = a_1$, $c_2 = \frac{1}{2}(a_2 + a_1^2)$, ..., i.e., $y(x) = x^{a_0} + a_1 x^{a_0+1} + \frac{1}{2}(a_2 + a_1^2)x^{a_0+2} + \ldots$. Note that the assumed form of solution is natural since $z \equiv x^{a_0}$ satisfies the nearby equation $xz' = a_0 z$.

15 If

$$t^2 y' = \begin{pmatrix} -2t & 4 \\ -2t^2 & 5t \end{pmatrix} y,$$

the first row $t^2 \dot{y}_1 = -2ty_1 + 4y_2$ implies that $4y_2 = t^2 \dot{y}_1 + 2ty_1$, so $4\dot{y}_2 = t^2 \ddot{y}_1 + 4t\dot{y}_1 + 2y_1$. The second row $t\dot{y}_2 = -2ty_1 + 5y_2$ therefore reduces to the Cauchy-Euler (and separable first-order) equation $t^2\ddot{y}_1 = t\dot{y}_1$ for $\dot{y}_1$. Thus, $y_1(t) = c_1 + c_2t^2$, and this implies that $y_2(t) = \frac{c_1 t}{2} + c_2 t^3$. The matrix

$$\Phi(t) = \begin{pmatrix} 2 & t^2 \\ t & t^3 \end{pmatrix},$$

corresponding to the convenient choice $c = \binom{2}{1}$, satisfies the matrix equation. Note that it becomes singular when $t = 0$.

16 Since $y(x) = z(s)$ for $s = 1/x$ implies that $y' = -s^2\frac{dz}{ds}$ and $y'' = s^4\frac{d^2z}{ds^2} + 2s^3\frac{dz}{ds}$ and the transformed equation $s^2\frac{d^2z}{ds^2} + \frac{3}{2}s\frac{dz}{ds} + sz = 0$ has a regular singular point at $s = 0$, we find the general solution $z(s)$ for $s > 0$ as $z(s) = \alpha(1-\frac{2s}{5}+\frac{2}{15}s^2+\ldots)+\frac{\beta}{s^{1/2}}(1-2s+\frac{2}{3}s^2+\ldots)$. Thus, the general solution for y is $y(x) = \alpha\left(1 - \frac{2}{5x} + \frac{2}{15x^2} + \ldots\right) + \beta x^{1/2}\left(1 - \frac{2}{x} + \frac{2}{3x^2} + \ldots\right)$, where $\beta = \lim_{x\to\infty}(x^{-1/2}y(x))$ and $\alpha = \lim_{x\to\infty}(y - \beta x^{1/2})$ are specified by asymptotic limits as $x \to \infty$ (or, equivalently, initial conditions as $s \to 0^+$).

17 (a) If we seek solutions $y(x) = \sum_{n=0}^{\infty} c_n x^{n+r}$, $c_0 \neq 0$, of Bessel's equation $x^2y'' + xy' + (x^2 - \frac{9}{4})y = 0$ of order $\frac{3}{2}$, we need $r = \pm 3/2$. The series $y_1(x) = \sum_{n=0}^{\infty} c_{n1}x^{n+3/2}$ will require $4c_{11}x^{5/2} + \sum_{n=2}^{\infty}[n(n+3)c_{n1} + c_{n-2\,1}]x^{n+3/2} = 0$, so $y_1(x) = c_{01}x^{3/2}(1 - \frac{1}{10}x^2 + \frac{1}{280}x^4 - \ldots)$. Likewise, $y_2(x) = \frac{c_{02}}{x^{3/2}}(1 + \frac{1}{2}x^2 - \frac{1}{8}x^4 + \ldots)$.

(b) The recurrence relations $J_{\pm\frac{3}{2}}(x) = \pm\frac{1}{x}J_{\pm\frac{1}{2}}(x) - J_{\mp\frac{1}{2}}(x)$ and the representations $J_{\frac{1}{2}}(x) = \sqrt{\frac{2}{\pi x}}\sin x$ and $J_{-\frac{1}{2}}(x) = \sqrt{\frac{2}{\pi x}}\cos x$ provide $J_{\frac{3}{2}}(x) = \sqrt{\frac{2}{\pi x^3}}(\sin x - x\cos x)$ and $J_{-\frac{3}{2}}(x) = -\sqrt{\frac{2}{\pi x^3}}(\cos x + x\sin x)$, which agree with the series obtained for y_1 and y_2 with $c_{01} = \frac{1}{3}\sqrt{\frac{2}{\pi}}$ and $c_{02} = -\sqrt{\frac{2}{\pi}}$.

18 Using the power series for $e^{\frac{xt}{2}}$ and $e^{-\frac{x}{2t}}$, we get $e^{\frac{xt}{2}}e^{-\frac{x}{2t}} = \sum_{k=0}^{\infty}\frac{1}{k!}\left(\frac{xt}{2}\right)^k \sum_{m=0}^{\infty}\frac{1}{m!}\left(-\frac{x}{2t}\right)^m$. If we let $a_k = \frac{1}{k!}\left(\frac{x}{2}\right)^k$ and $b_m = \frac{1}{m!}\left(-\frac{x}{2}\right)^m$, the Cauchy product formula implies that

$$e^{\frac{x}{2}(t-\frac{1}{t})} = \sum_{n=-\infty}^{\infty} t^n \sum_{m=0}^{\infty} a_{n+m}b_m = \sum_{n=-\infty}^{\infty} t^n \sum_{m=0}^{\infty} \frac{(-1)^m x^{n+2m}}{2^{n+2m}(n+m)!m!}$$

$$= \sum_{n=-\infty}^{\infty} t^n J_n(x).$$

Since $J_{-m}(x) = (-1)^m J_m(x)$, we obtain $e^{\frac{x}{2}(t-\frac{1}{t})} = J_0(x) + \sum_{n=1}^{\infty} J_n(x)(t^n + (-t)^{-n})$.

19 Since $x^2 J_0'' + x J_0' + x^2 J_0 = 0$, $x J_0 = -x J_0'' - J_0'$. Differentiating this result implies that $x J_0''' + 2J_0'' + x J_0' + J_0 = 0$. Multiplying by x and substituting for $x J_0$, we find that $x^2 J_0''' + x J_0'' + (x^2 - 1)J_0' = 0$, i.e., J_0' satisfies Bessel's equation of order one.

20 If $y(x) = z(s)$, then $y'(x) = -s^2 \frac{dz}{ds}$ and $y''(x) = s^4 \frac{d^2 z}{ds^2} + 2s^3 \frac{dz}{ds}$ for $s = 1/x$. Thus, $y'' + 2\left(\frac{1}{x^3} + \frac{1}{x}\right) y' + \frac{\lambda}{x^4} y = 0$ implies the Hermite equation $\frac{d^2 z}{ds^2} - 2s\frac{dz}{ds} + \lambda z = 0$ for $z(s)$. If we seek a series solution $z(s) = \sum_{n=0}^{\infty} c_n s^n$, we obtain

$$z(s) = \alpha \left(1 - \frac{\lambda}{2} s^2 + \frac{(4-\lambda)(-\lambda)}{24} s^4 + \dots\right)$$
$$+ \beta s \left(1 + \frac{(2-\lambda)}{6} s^2 + \frac{(6-\lambda)(2-\lambda)}{120} s^4 + \dots\right)$$

near $s = 0$ and, thereby, the approximation

$$y(x) = \alpha \left(1 - \frac{\lambda}{2x^2} + \frac{(4-\lambda)(-\lambda)}{24x^4} + \dots\right)$$
$$+ \frac{\beta}{x} \left(1 + \frac{(2-\lambda)}{6x^2} + \frac{(6-\lambda)(2-\lambda)}{120x^4} + \dots\right)$$

for large x, where $\alpha = \lim_{x\to\infty} y(x)$ and $\beta = \lim_{x\to\infty} x(y(x) - \alpha)$.

4

Systems of linear differential equations

4.1 The fundamental matrix

A *system* of n linear first-order differential equations in standard form consists of n coupled scalar equations

$$\begin{cases} \frac{dx_1}{dt} = a_{11}(t)x_1 + a_{12}(t)x_2 + \ldots + a_{1n}(t)x_n + f_1(t) \\ \frac{dx_2}{dt} = a_{21}(t)x_1 + a_{22}(t)x_2 + \ldots + a_{2n}(t)x_n + f_2(t) \\ \quad \vdots \\ \frac{dx_n}{dt} = a_{n1}(t)x_1 + a_{n2}(t)x_2 + \ldots + a_{nn}(t)x_n + f_n(t) \end{cases} \tag{4.1}$$

or, more simply, the vector first-order system

$$\dot{x} = A(t)x + f(t), \tag{4.2}$$

where

$$x = \begin{pmatrix} x_1 \\ \vdots \\ x_n \end{pmatrix}$$

is a column n-vector, as are $\dot{x}$ and f, and A is the $n \times n$ matrix

$$A(t) = \begin{pmatrix} a_{11}(t) & \ldots & a_{1n}(t) \\ a_{21}(t) & & a_{2n}(t) \\ \vdots & & \\ a_{n1}(t) & \ldots & a_{nn}(t) \end{pmatrix}$$

Note that (4.2) also makes sense when x and f are $n \times n$ matrices. (Readers needing to review practical aspects of linear algebra might consult Bronson (1970).)

The initial value problem, consisting of the system (4.2) with an arbitrary initial vector $x(t_0)$ prescribed at some point t_0 will have a unique solution $x(t)$ on any interval where the entries of $A(t)$ and $f(t)$ remain continuous. Moreover, we can always shift the t coordinate to get $t_0 = 0$. The homogeneous system, obtained when $f(t) \equiv 0$, will actually have n *linearly independent* vector solutions $\phi_i(t), i = 1, 2, \ldots, n$, corresponding to the n linearly independent choices for the initial n-vector $x(0)$; its general solution is given by the linear combination $\sum_{i=1}^{n} \phi_i(t) c_i$ for arbitrary constants c_i.

We will accept this theory without detailed proof and shall define an $n \times n$ *fundamental matrix* solution of the homogeneous system

$$\dot{\Phi} = A(t)\Phi$$

by setting

$$\Phi(t) \equiv (\phi_1(t) \;\; \phi_2(t) \ldots \phi_n(t))$$

since the usual rules for vector-matrix multiplication imply that the n column vectors ϕ_i of Φ satisfy $\dot{\phi}_i(t) = A(t)\phi_i(t)$. Because $\Phi(t)$ is invertible when its columns are linearly independent, the unique solution of the homogeneous vector initial value problem is readily confirmed to be

$$x(t) = \Phi(t)\Phi^{-1}(0)x(0).$$

(Check it!) The solution space for the homogeneous differential system $\dot{x} = A(t)x$ is then an n-dimensional linear vector space, with $\phi_1, \ldots, \phi_n$ as a basis whose linear combinations span all possible solutions, i.e., all vector solutions of $\dot{x} = A(t)x$ are of the form Φc for some vector

$$c = \Phi^{-1}(0)x(0) = \begin{pmatrix} c_1 \\ c_2 \\ \vdots \\ c_n \end{pmatrix}.$$

Note that we can use the *Wronskian* determinant

$$W(t) \equiv |\Phi(t)|$$

to check the linear independence of the columns of any given solution matrix $\Phi(t)$ for the homogeneous system $\dot{\Phi} = A(t)\Phi$. (We concede that the actual evaluation of the determinant (when n is large) is nontrivial to carry out.) If $W(t) \neq 0$, we have n linearly independent solutions ϕ_i, so we have determined the unique solution of the initial value problem. Otherwise, when $W(t) = 0$, we must look for more linearly independent vector solutions of $\dot{\phi} = A(t)\phi$ to form a nonsingular $n \times n$ fundamental matrix Φ. (Complications are naturally expected at singular points of the equation where $A(t)$ may be undefined.) Note that linear dependence of the solutions ϕ_i at some t_0 implies linear dependence for all A values since any combination $\sum_{i=1}^{n} C_i\phi_i(t)$ of solutions which is zero at A_0 must be zero for all times t, by uniqueness. Given a fundamental matrix Φ, a variation of parameters technique will be developed to provide all solutions for the corresponding nonhomogeneous vector system (4.2). When $A(t)$ and $f(t)$ are only piecewise continuous, one must attempt to define a solution by patching solutions together at points of discontinuity.

Example 1 The two-dimensional system

$$\begin{cases} \dot{x}_1 = x_2 \\ \dot{x}_2 = tx_2 \end{cases}$$

of scalar equations can be written as the homogeneous vector system $\dot{x} = A(t)x$ with

$$x = \begin{pmatrix} x_1 \\ x_2 \end{pmatrix} \quad \text{and} \quad A(t) = \begin{pmatrix} 0 & 1 \\ 0 & t \end{pmatrix}.$$

The system is easily solved since we can immediately integrate the second scalar equation to get

$$x_2(t) = e^{t^2/2}x_2(0),$$

leaving the scalar equation $\dot{x}_1 = e^{t^2/2}x_2(0)$, which integrates to provide

$$x_1(t) = x_1(0) + \left(\int_0^t e^{s^2/2}ds\right)x_2(0).$$

The vector solution can be conveniently written as the linear combination

$$x(t) = \phi_1(t)x_1(0) + \phi_2(t)x_2(0),$$

using the two linearly independent vector solutions

$$\phi_1(t) = \begin{pmatrix} 1 \\ 0 \end{pmatrix} \quad \text{and} \quad \phi_2(t) = \begin{pmatrix} \int_0^t e^{s^2/2}ds \\ e^{t^2/2} \end{pmatrix},$$

which define the nonsingular matrix solution

$$\Phi(t) = \begin{pmatrix} 1 & \int_0^t e^{s^2/2}ds \\ 0 & e^{t^2/2} \end{pmatrix}$$

of $\dot{\Phi} = A(t)\Phi$ for all t with the nonzero Wronskian $e^{t^2/2}$. Because this fundamental matrix Φ has the initial value $\Phi(0) = I$, any vector solution of the system has the form $x(t) = \Phi(t)x(0)$.

Let us temporarily restrict attention to autonomous homogeneous systems

$$\dot{x} = Ax, \tag{4.3}$$

where A is a constant state matrix. Because (4.3) is *translation-invariant* with respect to t (i.e., the differential equation is unchanged by a shift of the independent variable from t to $s = t - t_0$ for any t_0), taking $t = 0$ as the initial point is generic. We will ultimately determine a special fundamental matrix solution of (4.3), which we will appropriately denote by e^{At}. First, however, let us (somewhat boldly) seek nontrivial vector solutions of (4.3) in the exponential form

$$x(t) = e^{\lambda t}v, \tag{4.4}$$

where the scalar λ and the vector v are, respectively, selected to be an eigenpair [i.e., corresponding eigenvalue and (nonzero) eigenvectors] of A defined by $Av = \lambda v$. Substituting (4.4) into (4.3), we have determined a solution, when $\dot{x} = \lambda e^{\lambda t}v = Ax = Ae^{\lambda t}v$. Because $e^{\lambda t} \neq 0, \lambda$ and v must satisfy the eigenvalue equation

$$(A - \lambda I)v = 0. \tag{4.5}$$

To get a nontrivial vector solution v, the matrix $A - \lambda I$ must be singular, so λ must satisfy the *characteristic polynomial*

$$p(\lambda) \equiv |A - \lambda I| \equiv \det(A - \lambda I) = 0, \tag{4.6}$$

a scalar polynomial in λ of degree n. Then, the eigenvector v is determined up to a scalar multiplier. (Introducing the ansatz (4.4), then, is completely analogous to seeking solutions $e^{\lambda t}$ of scalar constant-coefficient linear equations when λ is a root of the associated characteristic polynomial.) If the state matrix A has n linearly independent eigenvectors v_j, corresponding, respectively, to eigenvalues λ_j, we will thereby get n linearly independent solutions

$$\phi_i(t) = e^{\lambda_i t}v_i, \quad i = 1, 2, \ldots, n$$

and a corresponding nonsingular fundamental matrix $\Phi(t)$ with these vectors $\phi_i(t)$ as its n columns. (Moreover, the linear independence of the solutions to (4.3) at $t = 0$ implies the same for all t and vice versa.) Having a full set of n linearly independent eigenvectors v_j of A is definitely guaranteed by the usual theory when A has n distinct eigenvalues. If some eigenvalues are repeated and the full set of eigenvectors does not span an n-dimensional space, we will generally need to use *generalized eigenvectors* of A to get the general solution (i.e., a complete set of solutions) of the system (4.3).

Example 2 Consider the system $\dot{x} = Ax$ for $A = \begin{pmatrix} -1 & 6 \\ 1 & -2 \end{pmatrix}$. Here, A has the characteristic polynomial

$$|A - \lambda I| = \begin{vmatrix} -1-\lambda & 6 \\ 1 & -2-\lambda \end{vmatrix}$$
$$= (-1-\lambda)(-2-\lambda) - 6 = \lambda^2 + 3\lambda - 4 = (\lambda - 1)(\lambda + 4) = 0.$$

Eigenvectors v_1 and v_2, corresponding to the respective eigenvalues $\lambda_1 = 1$ and $\lambda_2 = -4$, must be nontrivial solutions of the algebraic equations

$$(A - I)v_1 = \begin{pmatrix} -2 & 6 \\ 1 & -3 \end{pmatrix} v_1 = 0 \text{ and } (A + 4I)v_2 = \begin{pmatrix} 3 & 6 \\ 1 & 2 \end{pmatrix} v_2 = 0.$$

Selecting

$$v_1 = \begin{pmatrix} 3 \\ 1 \end{pmatrix} \text{ and } v_2 = \begin{pmatrix} 2 \\ -1 \end{pmatrix}$$

provides us the linearly independent solutions

$$\phi_1(t) = \begin{pmatrix} 3 \\ 1 \end{pmatrix} e^t \text{ and } \phi_2(t) = \begin{pmatrix} 2 \\ -1 \end{pmatrix} e^{-4t}$$

of $\dot{x} = Ax$ and, thereby, the corresponding fundamental matrix

$$\Phi(t) = \begin{pmatrix} 3e^t & 2e^{-4t} \\ e^t & -e^{-4t} \end{pmatrix}.$$

One can directly check that $\dot{\Phi}(t) = A\Phi(t)$, that $\Phi(t)$ is nonsingular for all t since its determinant is $-5e^{-3t}$, and that the product

$$\Phi(t)c$$

for an arbitrary constant vector $c \equiv \begin{pmatrix} c_1 \\ c_2 \end{pmatrix}$ is the general vector solution of $\dot{x} = Ax$. Since

$$\Phi^{-1}(0) = \frac{1}{5}\begin{pmatrix} 1 & 2 \\ 1 & -3 \end{pmatrix},$$

the solution of the initial value problem $\dot{x} = Ax$ for any given $x(0)$ is

$$x(t) = \Phi(t)\Phi^{-1}(0)x(0) = \frac{1}{5}\begin{pmatrix} 3e^t & 2e^{-4t} \\ e^t & -e^{-4t} \end{pmatrix}\begin{pmatrix} 1 & 2 \\ 1 & -3 \end{pmatrix} x(0),$$

i.e., $x(t) \equiv e^{At}x(0)$ defines the *matrix exponential*

$$e^{At} \equiv \frac{1}{5}\begin{pmatrix} 3e^t + 2e^{-4t} & 6(e^t - e^{-4t}) \\ e^t - e^{-4t} & 2e^t + 3e^{-4t} \end{pmatrix}$$

for the given constant matrix A. Note that this e^{At} is nonsingular for all t and satisfies both $\frac{d}{dt}e^{At} = Ae^{At}$ and $e^{A0} = I$.

We will often have to explicitly find the inverse of a matrix. Many techniques, including row-reduction, are available. For 2×2 nonsingular matrices

$$M = \begin{pmatrix} a & b \\ c & d \end{pmatrix},$$

it is simple to check that the unique inverse is

$$M^{-1} = \frac{1}{ad - bc}\begin{pmatrix} d & -b \\ -c & a \end{pmatrix}.$$

Although we use two-dimensional problems for simplicity, the reason to introduce matrix methods is to solve higher-dimensional problems efficiently.

Example 3 Consider the automonous vector system $\dot{x} = Ax$ for $A = \begin{pmatrix} 4 & 1 \\ -8 & 8 \end{pmatrix}$. Here

$$\det(A - \lambda I) = \begin{vmatrix} 4 - \lambda & 1 \\ -8 & 8 - \lambda \end{vmatrix} = \lambda^2 - 12\lambda + 40 = (\lambda - 6)^2 + 4 = 0$$

implies that the eigenvalues $\lambda_{1,2} = 6 \pm 2i$ of A are complex conjugates. The equation $(A - \lambda_{1,2}I)v_{1,2} = 0$ will provide corresponding complex conjugate eigenvectors $v_{1,2}$. Thus, we (nonuniquely) solve

$$\begin{pmatrix} -2 - 2i & 1 \\ -8 & 2 - 2i \end{pmatrix} v_1 = 0$$

for $v_1 = \begin{pmatrix} 1 \\ 2 + 2i \end{pmatrix}$ and, using Euler's formula, we obtain the complex conjugate solutions

$$\phi_1(t) \equiv e^{(6+2i)t}\begin{pmatrix} 1 \\ 2+2i \end{pmatrix} = e^{6t}\begin{pmatrix} \cos 2t + i\sin 2t \\ 2(\cos 2t - \sin 2t) + 2i(\cos 2t + \sin 2t) \end{pmatrix}$$

and $\phi_2(t) = \overline{\phi_1(t)}$ to the given real differential equation. [One can check that the complex solution matrix $(\phi_1(t)\ \phi_2(t))$ has a nonzero (Wronskian) determinant.] The real and imaginary parts of $\phi_1(t)$, however, conveniently provide two linearly independent real solutions of $\dot{x} = Ax$ and so define the real fundamental matrix

$$\Psi(t) = e^{6t}\begin{pmatrix} \cos 2t & \sin 2t \\ 2(\cos 2t - \sin 2t) & 2(\cos 2t + \sin 2t) \end{pmatrix}.$$

(Readers should check that $\dot{\Psi} = A\Psi$ and that Ψ is nonsingular.) Since

$$\Psi(0) = \begin{pmatrix} 1 & 0 \\ 2 & 2 \end{pmatrix} \quad \text{and} \quad \Psi^{-1}(0) = \frac{1}{2}\begin{pmatrix} 2 & 0 \\ -2 & 1 \end{pmatrix},$$

we can define

$$e^{At} \equiv \Psi(t)\Psi^{-1}(0) = e^{6t}\begin{pmatrix} \cos 2t - \sin 2t & \frac{1}{2}\sin 2t \\ -4\sin 2t & \cos 2t + \sin 2t \end{pmatrix}$$

to obtain the general solution $x(t) = e^{At}x(0)$ of $\dot{x} = Ax$ for the given matrix A.

We will now develop the variation of parameters formula to solve the general nonhomogeneous linear system

$$\dot{x} = A(t)x + f(t) \tag{4.2}$$

(corresponding to the n scalar equations (4.1)) on an interval t, where the elements of $A(t)$ and $f(t)$ all remain continuous. We will presume that a nonsingular matrix solution $\Phi(t)$ of the corresponding homogeneous system $\dot{x} = A(t)x$ is explicitly known. Noting that

$$x(t) = \Phi(t)c,$$

for an arbitrary constant vector c, provides the general solution of the homogeneous vector system $\dot{x} = A(t)x$, we will use the variation of constants idea and seek a solution of the nonhomogeneous system (4.2) in the form

$$x(t) = \Phi(t)v(t) \tag{4.7}$$

for a time-varying vector $v(t)$ to be determined. Substitution into (4.2) implies that

$$\dot{x} = \Phi\dot{v} + \dot{\Phi}v = \Phi\dot{v} + A(t)\Phi v = A(t)\Phi v + f(t)$$

(since $\dot{\Phi} = A\Phi$), so we must simply ask that v satisfy

$$\Phi\dot{v} = f(t).$$

Integrating $\dot{v}(t) = \Phi^{-1}(t)f(t)$ from any initial point t_0, we obtain

$$v(t) = v(t_0) + \int_{t_0}^{t} \Phi^{-1}(s)f(s)ds,$$

where $v(t_0) = \Phi^{-1}(t_0)x(t_0)$ follows from (4.7). Thus, we find the unique solution

$$x(t) = \Phi(t)\Phi^{-1}(t_0)x(t_0) + \int_{t_0}^{t} \Phi(t)\Phi^{-1}(s)f(s)ds \tag{4.8}$$

to (4.2) (which can be immediately verified by direct substitution into the differential equation). Note that

$$x_p(t) \equiv \int^{t} G(t,s)f(s)ds$$

defines particular solutions of (4.2) for the matrix kernel

$$G(t,s) \equiv \Phi(t)\Phi^{-1}(s).$$

We naturally call G, which is the unique solution of $\dot{G} = A(t)G$, $G(s) = I$ for some s, a Green's function for the vector or matrix initial value problem, because it provides a direct link (through integration) between a given forcing function f and corresponding particular solutions x_p. That the general solution of (4.2) is the sum of a solution of the homogeneous system and a particular solution is exactly the same affine structure found for higher-order scalar equations. The variation of parameters formula is now, however, somewhat more attractive than before due to the streamlined matrix notation.

In the special case that A is constant, it is convenient to rewrite the Green's function $G(t,s) = e^{At}e^{-As}$ as $G(t,s) \equiv e^{A(t-s)}$ to get the even more attractive formula

$$x(t) = e^{A(t-t_0)}x(t_0) + \int_{t_0}^{t} e^{A(t-s)}f(s)ds, \tag{4.9}$$

which directly generalizes both the solution obtained in Chapter 1 for the scalar first-order linear equation $\dot{x} = ax + f(t)$ to the vector or matrix equation and the solution for certain second-order equations given by example 6 in Chapter 2.

As the exercises will demonstrate, a method of undetermined coefficients for systems (4.2) also results when f is a linear combination of polynomials times exponentials.

In applications, we often need to consider linear *boundary value problems*, rather than initial value problems, where solutions of the nonhomogeneous system (4.2) (on an interval $t_0 < t < t_1$, where $A(t)$ and $f(t)$ are assumed to be continuous) are required to satisfy linear auxiliary conditions commonly in the form

$$B_0 x(t_0) + B_1 x(t_1) = g \tag{4.10}$$

for prescribed constant $n \times n$ matrices B_1 and B_2 and a given constant n-vector g. This, of course, is an initial value problem when, for example, B_0 is nonsingular and $B_1 \equiv 0$. Since any solution of (4.2) must have the form (4.8) for some unknown initial vector $x(t_0)$, the linear boundary condition (4.10) requires $x(t_0)$ to satisfy the linear algebraic system

$$B_0 x(t_0) + B_1 \Phi(t_1) \left[\Phi^{-1}(t_0) x(t_0) + \int_{t_0}^{t_1} \Phi^{-1}(s) f(s) ds \right] = g$$

or

$$\mathcal{B} \Phi^{-1}(t_0) x(t_0) = \tilde{g} \quad \text{for} \quad \mathcal{B} \equiv B_0 \Phi(t_0) + B_1 \Phi(t_1)$$

and $\tilde{g} \equiv g - B_1 \Phi(t_1) \int_{t_0}^{t_1} \Phi^{-1}(s) f(s) ds$. This system of linear algebraic equations for $x(t_0)$ will have a unique solution

$$x(t_0) = (\mathcal{B}(\Phi^{-1}(t_0))^{-1} \tilde{g} = \Phi(t_0) \mathcal{B}^{-1} \tilde{g},$$

provided the coefficient matrix $\mathcal{B}$ is nonsingular.

Thus, the unique solution of the boundary value problem [(4.2)–(4.10)] can be conveniently rewritten in the compact form

$$x(t) = \Phi(t) B_g^{-1} \int_{t_0}^{t_1} G(t, s) f(s) ds$$

for the matrix Green's function

$$G(t, s) \equiv \begin{cases} \Phi(t) \mathcal{B}^{-1} B_0 \Phi(t_0) \Phi^{-1}(s), & s \leq t \\ -\Phi(t) \mathcal{B}^{-1} B_1 \Phi(t_1) \Phi^{-1}(s), & s > t. \end{cases}$$

When B is singular, the *Fredholm alternative* for linear algebraic systems implies that the boundary value problem has no solution unless $\tilde{g}$ happens to lie in the range of $\mathcal{B}$, so an appropriate $x(t_0)$ can be found. Then, the number of solutions to the boundary value problem is determined by the *rank* of the matrix $\mathcal{B}$. Note that determining an initial vector $x(t_0)$, appropriate to the boundary condition (4.10), is the idea behind numerical *shooting methods*, where guesses for $x(t_0)$ are iteratively made to sequentially minimize the discrepancy $B_0 x(t_0) + B_1 x(t_1) - g$ in solving (4.10) for the corresponding solution (4.8) of the linear differential system (4.2).

Example 4 Consider the nonhomogeneous problem

$$\dot{x} = \begin{pmatrix} 4 & 2 \\ 3 & 3 \end{pmatrix} x + \begin{pmatrix} e^t \\ e^{2t} \end{pmatrix}, \; x(0) = \begin{pmatrix} 1 \\ 0 \end{pmatrix}.$$

Since the constant state matrix $A = \begin{pmatrix} 4 & 2 \\ 3 & 3 \end{pmatrix}$ has the eigenpairs $\left(1, \begin{pmatrix} 2 \\ -3 \end{pmatrix}\right)$ and $\left(6, \begin{pmatrix} 1 \\ 1 \end{pmatrix}\right)$, the homogeneous system $\dot{x} = Ax$ has the fundamental matrix

$$\Phi(t) = \begin{pmatrix} 2e^t & e^{6t} \\ -3e^t & e^{6t} \end{pmatrix}.$$

Using its inverse

$$\Phi^{-1}(t) = \frac{1}{5}\begin{pmatrix} e^{-t} & -e^{-t} \\ 3e^{-6t} & 2e^{-6t} \end{pmatrix}$$

allows us to use the variation of constants formula (4.8) to provide the unique solution

$$\begin{aligned} x(t) &= \Phi(t)\left[\Phi^{-1}(0)\begin{pmatrix} 1 \\ 0 \end{pmatrix} + \int_0^t \Phi^{-1}(s)\begin{pmatrix} e^s \\ e^{2s} \end{pmatrix} ds\right] \\ &= \frac{1}{5}\begin{pmatrix} 2te^t + \frac{17}{5}e^t - \frac{5}{2}e^{2t} + \frac{41}{10}e^{6t} \\ -3te^t - \frac{33}{5}e^t + \frac{5}{2}e^{2t} + \frac{41}{10}e^{6t} \end{pmatrix} \end{aligned}$$

of the given initial value problem (as can be directly checked). The same result can also be obtained by introducing

$$e^{At} = \Phi(t)\Phi^{-1}(0) = \frac{1}{5}\begin{pmatrix} 2e^t + 3e^{6t} & -2(e^t - e^{6t}) \\ -3(e^t - e^{6t}) & 3e^t + 2e^{6t} \end{pmatrix}$$

and using (4.9). The manipulations needed to get the correct answer must be carefully carried out stepwise. Readers are therefore urged to do so for themselves and to employ Maple or Mathematica for corresponding higher-dimensional problems, where the possibility for careless error is increased.

By contrast, if we allow the form of the forcing term $\begin{pmatrix} e^t \\ e^{2t} \end{pmatrix}$ to suggest a particular solution

$$x_p(t) = (M + Nt)e^t + Pe^{2t}$$

for constant vectors M, N, and P to be determined, we would have

$$\dot{x}_p = (M + N + Nt)e^t + 2Pe^{2t}.$$

(Note that a preliminary attempt with $N = 0$ would fail, since $\lambda = 1$ is an eigenvalue of A and multiples of e^t already appear in the homogeneous solution.) Thus, the differential equation will require

$$(M + N + Nt)e^t + 2Pe^{2t} = A[(M + Nt)e^t + Pe^{2t}] + \begin{pmatrix} e^t \\ e^{2t} \end{pmatrix},$$

and the linear independence of e^t, te^t, and e^{2t} forces M, N, and P to satisfy the linear system

$$\begin{cases} M + N = AM + \begin{pmatrix} 1 \\ 0 \end{pmatrix} \\ N = AN \\ \text{and} \\ 2P = AP + \begin{pmatrix} 0 \\ 1 \end{pmatrix}. \end{cases}$$

Because $A - 2I$ is nonsingular, we uniquely obtain

$$P = -(A - 2I)^{-1}\begin{pmatrix} 0 \\ 1 \end{pmatrix} = \frac{1}{4}\begin{pmatrix} 1 & -2 \\ -3 & 2 \end{pmatrix}\begin{pmatrix} 0 \\ 1 \end{pmatrix} = -\frac{1}{2}\begin{pmatrix} 1 \\ -1 \end{pmatrix}.$$

Since N must be an eigenvector of A corresponding to the eigenvalue 1, we set $N = \begin{pmatrix} 2 \\ -3 \end{pmatrix} c_0$ for some constant c_0. Then, M must satisfy

$$(A - I)M = \begin{pmatrix} 3 & 2 \\ 3 & 2 \end{pmatrix} M = \begin{pmatrix} 2c_0 - 1 \\ -3c_0 \end{pmatrix}.$$

To have a solution of this nonhomogeneous system requires the right-hand side to be in the one-dimensional range of $A - I$, spanned by the vector $\begin{pmatrix} 1 \\ 1 \end{pmatrix}$, so $2c_0 - 1 = -3c_0$ or $c_0 = \frac{1}{5}$. Then, M will be determined up to arbitrary multiples of this eigenvector $\begin{pmatrix} 2 \\ -3 \end{pmatrix}$. We will simply take $M = \frac{1}{5}\begin{pmatrix} 1 \\ 0 \end{pmatrix}$ to get the particular solution

$$x_p(t) = \frac{1}{5}\begin{pmatrix} 1+2t \\ -3t \end{pmatrix} e^t - \frac{1}{2}\begin{pmatrix} 1 \\ -1 \end{pmatrix} e^{2t}.$$

Since any solution of the linear system has the affine form

$$x(t) = \Phi(t)k + x_p(t)$$

for some constant vector k, the initial condition $x(0) = \begin{pmatrix} 1 \\ 0 \end{pmatrix} = \Phi(0)k$ $+x_p(0)$ uniquely determines the vector k and the solution already found through variation of constants.

Example 5 The matrix $A = \begin{pmatrix} 7 & -1 \\ 9 & 1 \end{pmatrix}$ has the characteristic polynomial $|A - \lambda I| = \begin{vmatrix} 7-\lambda & -1 \\ 9 & 1-\lambda \end{vmatrix} = (\lambda - 4)^2 = 0$. Moreover, there is only one linearly independent eigenvector $\begin{pmatrix} 1 \\ 3 \end{pmatrix}$ corresponding to the repeated eigenvalue $\lambda = 4$. We are therefore guaranteed that

$$\phi_1(t) = \begin{pmatrix} 1 \\ 3 \end{pmatrix} e^{4t}$$

is a vector solution of $\dot{x} = Ax$, and we might attempt to mimic the familiar reduction of order procedure by seeking a second linearly independent solution as

$$\phi_2(t) = e^{4t} w(t)$$

for some variable vector w. The differential equation $\dot{\phi}_2 = A\phi_2$ then requires w to satisfy the system

$$\dot{w} = (A - 4I)w,$$

whose state matrix $A - 4I$ has zero as a double eigenvalue. Our experience with scalar equations and our success with power series both suggest that we might first seek solutions $w(t)$ as a Maclaurin series

$$w(t) = b_0 + b_1 t + b_2 t^2 + b_3 t^3 + \dots$$

with vector coefficients b_j. Since $\dot{w}(t) = b_1 + 2b_2 t + 3b_3 t^2 + \dots$, we will need

$$b_1 + 2b_2 t + 3b_3 t^2 + \dots = (A - 4I)b_0 + (A - 4I)b_1 t + (A - 4I)b_2 t^2 + \dots.$$

Equating coefficients of successive powers of t, we ask that

$$b_1 = (A - 4I)b_0, \quad b_2 = \frac{1}{2}(A - 4I)b_1$$
$$= \frac{1}{2}(A - 4I)^2 b_0, \quad b_3 = \frac{1}{3}(A - 4I)b_2 = \frac{1}{3!}(A - 4I)^3 b_0, \text{ etc.}$$

But $(A - 4I)^2 = 0$ (check it!). This implies that b_2 and later coefficients b_j in the series for $w(t)$ are all trivial. Moreover,

$$(A - 4I)b_1 = (A - 4I)^2 b_0 = 0$$

implies that b_1 must be an eigenvector of A (corresponding to the repeated eigenvalue) and b_0 must be a corresponding *generalized eigenvector* that satisfies

$$(A - 4I)b_0 = b_1.$$

If we conveniently select b_1 to be $\begin{pmatrix} 1 \\ 3 \end{pmatrix}$, b_0 must satisfy the singular system $(A - 4I)b_0 = \begin{pmatrix} 1 \\ 3 \end{pmatrix}$. Its solutions are $b_0 = \begin{pmatrix} 1 \\ 2 \end{pmatrix} + \begin{pmatrix} 1 \\ 3 \end{pmatrix} \alpha_0$ for an arbitrary constant α_0. Taking $\alpha_0 = 0$, for simplicity, finally provides the vector $w(t)$ as a linear function of t, the desired solution

$$\phi_2(t) = e^{4t} w(t) = e^{4t} \begin{pmatrix} 1 + t \\ 2 + 3t \end{pmatrix},$$

and thereby the fundamental matrix

$$\Phi(t) = (\phi_1 \, \phi_2) = e^{4t} \begin{pmatrix} 1 & 1 + t \\ 3 & 2 + 3t \end{pmatrix}.$$

(The reader should directly check that Φ is a nonsingular solution of $\dot{\Phi} = A\Phi$. She or he shouldn't be surprised to find the fundamental matrix as a polynomial multiple of e^{4t}.)

Before proceeding, we observe that it was not accidental that the constant matrix A in example 5 satisfied $(A - 4I)^2 = 0$, since the *Cayley-Hamilton theorem* guarantees that every square matrix A satisfies its own characteristic polynomial $p(\lambda) \equiv \det(A - \lambda I) = 0$, i.e., $p(A) = 0$. The series method generalizes, allowing us to obtain m_k linearly independent vector solutions ϕ of any autonomous homogeneous system $\dot{x} = Ax$ in the form $\phi(t) = e^{\lambda_k t} q_k(t)$, using a polynomial q_k in t of degree $m_k - 1$ or less with unknown vector coefficients, whenever λ_k is a root of the characteristic polynomial of degree m_k, i.e., $(\lambda - \lambda_k)^{m_k}$ is a factor of $p(\lambda)$. If $p(\lambda) = 0$ has r distinct roots λ_k, with corresponding multiplicities m_k summing to n, we will be able to generate an n-dimensional basis of solutions for $\dot{x} = Ax$. To get the m_k solutions corresponding to λ_k, we shall set

$$\phi(t) = e^{\lambda_k t} q_k(t) \equiv e^{\lambda_k t} \sum_{j=0}^{m_k - 1} \frac{c_{jk} t^j}{j!},$$

expecting the possibility that later terms c_{jk} in the sum may all be zero. Unless λ_k is purely imaginary, note that these solutions will grow or decay, roughly like the exponential $e^{\lambda_k t}$, as $t \to \infty$. Substitution of ϕ into the differential equation $\dot{x} = Ax$ and matching coefficients of powers t^j in the usual manner will successively require that

$$(A - \lambda_k I) c_{jk} = c_{j+1k},$$

for each j. We will proceed optimistically by first attempting to get solutions with (simply) the vector $q_k(t) = c_{0k}$ satisfying the eigenvector equation $(A - \lambda_k I) c_{0k} = 0$. If this provides the needed m_k linearly independent solutions of $\dot{x} = Ax$ corresponding to λ_k, we shall quit. If not, we shall next seek solutions ϕ_k with $q_k(t) = c_{0k} + c_{1k} t$, where the c_{1k}s are eigenvectors satisfying $(A - \lambda_k I) c_{1k} = 0$ and the c_{0k}s are related *generalized eigenvectors* such that $(A - \lambda_k I) c_{0k} = c_{1k}$. If this still fails to provide a total of m_k linearly independent solutions ϕ_k, we would next seek additional solutions with $q_k(t) = c_{0k} + c_{1k} t + \frac{1}{2} c_{2k} t^2$, where $(A - \lambda_k I) c_{2k} = 0$, $(A - \lambda_k I) c_{1k} = c_{2k}$, and $(A - \lambda_k I) c_{0k} = c_{1k}$ nonuniquely define c_{2k} to be an eigenvector, with c_{1k} and c_{2k} creating a chain of generalized eigenvectors, which all necessarily linearly independent. Continuing in this fashion for $k = 1, 2, \ldots, r$, we will ultimately obtain a full set of n linearly independent solutions, which form a basis of the n-dimension solution space for $\dot{x} = Ax$. The more sophisticated readers will note the parallels between this (undetermined coefficients) procedure and that used to express any matrix in terms of a basis that puts the matrix into its *Jordan canonical form*. Others might note parallels to *normal mode analysis* as used, for example, for vibrating systems.

To illustrate the preceding search for solutions more concretely, consider the example

$$A = \begin{bmatrix} 0 & 1 & 0 \\ 0 & 0 & 1 \\ 0 & 0 & 0 \end{bmatrix},$$

which has 0 as an eigenvalue of multiplicity three. It has the eigenvector (nullvector) $e_1 = \begin{pmatrix} 1 \\ 0 \\ 0 \end{pmatrix}$, which is thereby a constant solution

$$\phi_1 = e_1$$

of $\dot{x} = Ax$. All other constant solutions are multiples of e_1, so we next seek solutions of the form

$$\phi = c_0 + c_1 t, \quad c_1 \neq 0.$$

Then, $\dot{\phi} = A\phi$ implies that $c_1 = A(c_0 + c_1 t)$, so we will need $Ac_1 = 0$ and $Ac_0 = c_1$. Picking $c_1 =$ the nullvector e_1 and the generalized nullvector $c_0 = e_2 = \begin{pmatrix} 0 \\ 1 \\ 0 \end{pmatrix}$ is satisfactory. (Nonzero multiples of e_1 could also be used, and multiples of e_1 could also be added to e_2.) Thus, we obtain a second linearly independent solution

$$\phi_2 = e_2 + e_1 t = \begin{pmatrix} t \\ 1 \\ 0 \end{pmatrix}$$

of $\dot{x} = Ax$. Finally, if we seek a solution

$$\phi_3 = c_0 + c_1 t + \frac{1}{2} c_2 t^2, c_2 \neq 0,$$

we will need $Ac_2 = 0$, $Ac_1 = c_2$, and $Ac_0 = c_1$. One solution is provided by $c_2 = e_1, c_1 = e_2$, and $c_3 = e_3 = \begin{pmatrix} 0 \\ 0 \\ 1 \end{pmatrix}$, i.e.,

$$\phi_3 = \begin{pmatrix} \frac{1}{2}t^2 \\ t \\ 1 \end{pmatrix}.$$

Thus, we have found the nonsingular matrix solution

$$\Phi \equiv \begin{pmatrix} 1 & t & \frac{t^2}{2} \\ 0 & 1 & t \\ 0 & 0 & 1 \end{pmatrix}$$

of the equation $\dot{x} = Ax$. Since $\Phi(0) = I$, it follows that $\Phi(t) = e^{At}$.

For the matrix

$$A = \begin{bmatrix} 0 & 0 & 0 \\ 0 & 0 & 1 \\ 0 & 0 & 0 \end{bmatrix},$$

the two nullvectors e_1 and e_2 provide linearly independent solutions. To get a third linearly independent solution

$$\phi = c_0 + c_1 t, \quad c_1 \neq 0,$$

we will need $Ac_1 = 0$ and $Ac_0 = c_1$. A solution is given by $c_1 = e_2$ and $c_0 = e_3$, i.e., $\phi = \begin{pmatrix} 0 \\ t \\ 1 \end{pmatrix}$. Thus, a nonsingular matrix solution of the new system $\dot{x} = Ax$ is given by

$$e^{At} = \begin{pmatrix} 1 & 0 & 0 \\ 0 & 1 & t \\ 0 & 0 & 1 \end{pmatrix}.$$

To continue to progress, we must generate successful methods to obtain fundamental matrices more generally, e.g., when the state matrix $A(t)$ in $\dot{x} = A(t)x$ is not constant. For difficult problems, combined asymptotic and numerical approximations may be necessary.

Example 6 Suppose

$$A(t) = \begin{pmatrix} a_1(t) & 0 & \dots & 0 \\ 0 & a_2(t) & \dots & 0 \\ & & \ddots & \\ 0 & 0 & \dots & a_n(t) \end{pmatrix}$$

is a *diagonal matrix*. Then, the system $\dot{x} = A(t)x$ is equivalent to the n decoupled scalar equations $\dot{x}_i = a_i(t)x_i$ and the general solution $x(t)$ is determined by its components

$$x_i(t) = e^{\int_0^t a_i(s)ds} x_i(0), \; i = 1, \dots, n.$$

We naturally write the fundamental matrix for the vector system as the diagonal matrix

$$\Phi(t) = \begin{pmatrix} e^{\int_0^t a_1(s)ds} & & \\ & \ddots & \\ & & e^{\int_0^t a_n(s)ds} \end{pmatrix}$$

or we, more simply, use the suggestive notation $\Phi(t) \equiv \exp[\int_0^t A(s)ds]$. More generally, if $A(t)$ is a block-diagonal matrix, with $m_i \times m_i$ matrices $a_i(t)$ along its diagonal (and trivial blocks elsewhere), the fundamental matrix $\Phi(t)$ will also be block-diagonal with $m_i \times m_i$ diagonal matrix entries $\Phi_i(t) \equiv \exp[\int_0^t a_i(s)ds]$, defined by $\dot{\Phi}_i = a_i(t)\Phi_i$, $\Phi_i(0) = I_{m_i}$. One might even find it convenient to define more general functions of diagonal matrices.

Example 7 Let us generalize example 6. Suppose

$$A(t) = \begin{pmatrix} B(t) & 0 \\ C(t) & D(t) \end{pmatrix}$$

is *block-triangular*, with square diagonal elements B and D (not necessarily of the same size), and suppose fundamental matrices $\Phi_1(t)$ and $\Phi_2(t)$, respectively, are known for both linear systems

$$\dot{z}_1 = B(t)z_1 \quad \text{and} \quad \dot{z}_2 = D(t)z_2.$$

Then, we can solve the linear system $\dot{x} = A(t)x$ by introducing appropriately compatible subvectors $\begin{pmatrix} x_1 \\ x_2 \end{pmatrix} = x$ and rewriting the original system as

$$\begin{cases} \dot{x}_1 = B(t)x_1 \\ \dot{x}_2 = C(t)x_1 + D(t)x_2 \end{cases}.$$

Integration immediately yields $x_1(t) = \Phi_1(t)\Phi_1^{-1}(0)x_1(0)$, so substitution into the second system, followed by variation of parameters, yields

$$x_2(t) = \Phi_2(t)\left[\Phi_2^{-1}(0)x_2(0) + \left(\int_0^t \Phi_2^{-1}(s)C(s)\Phi_1(s)ds\right)\Phi_1^{-1}(0)x_1(0)\right]$$

$$\equiv \Phi_2(t)\Phi_2^{-1}(0)x_2(0) + \Phi_3(t)\Phi_1^{-1}(0)x_1(0),$$

defining $\Phi_3(t)$. Thus, the homogeneous block-triangular system $\dot{x} = A(t)x$ has a fundamental matrix

$$\Phi(t) = \begin{bmatrix} \Phi_1(t) & 0 \\ \Phi_3(t) & \Phi_2(t) \end{bmatrix}.$$

Readers should directly check that $\Phi(t)$ is a nonsingular solution of $\dot{\Phi} = A(t)\Phi$.

Note further that the fundamental matrix $\Phi(t)$ has the inverse

$$\Phi^{-1}(t) = \begin{bmatrix} \Phi_1^{-1}(t) & 0 \\ -\Phi_2^{-1}(t)\Phi_3(t)\Phi_1^{-1}(t) & \Phi_2^{-1}(t) \end{bmatrix}.$$

(Check that $\Phi\Phi^{-1} = I$.). Since the nonhomogeneous system

$$\dot{x} = A(t)x + f$$

has the particular solution $x_p(t) = \Phi(t)\int^t \Phi^{-1}(s)f(s)ds$, if we split $x = \begin{pmatrix} x_1 \\ x_2 \end{pmatrix}$ and $f = \begin{pmatrix} f_1 \\ f_2 \end{pmatrix}$ as we did Φ, we then get a particular solution with component subvectors

$$x_{1\,p}(t) = \Phi_1(t)\int^t \Phi_1^{-1}(s)f_1(s)ds$$

and

$$\begin{aligned} x_{2\,p}(t) = {} & \Phi_3(t)\int^t \Phi_1^{-1}(s)f_1(s)ds \\ & + \Phi_2(t)\int^t \Phi_2^{-1}(s)\left[-\Phi_3(s)\Phi_1^{-1}(s)f_1(s) + f_2(s)\right] ds. \end{aligned}$$

The reader should check that this solution would also result if variation of parameters were used on the nonhomogeneous system for x_1 and, again, on the resulting nonhomogeneous system for x_2.

The technique presented for example 7 is, of course, a direct generalization of the usual method to solve the triangular system of scalar equations

$$\begin{cases} \dot{x}_1 = b_1(t)x_1, \\ \dot{x}_2 = c_1(t)x_1 + b_2(t)x_2, \\ \dot{x}_3 = c_2(t)x_1 + c_3(t)x_2 + b_3(t)x_3 \end{cases}$$

by integrating the first equation to get x_1, substituting that x_1 into the second equation and integrating to get x_2, and finally substituting x_1 and x_2 into the third equation to get x_3 (cf. example 1). That, in turn, mimics Gaussian elimination, as traditionally used in linear algebra.

The analogous approach can be used for upper triangular systems, but no universal technique is available for

$$\dot{x} = A(t)x \tag{4.12}$$

with a general state matrix $A(t)$. It is, nonetheless, useful to develop the following reduction of order method for (4.12). Suppose we know r linearly independent vector solutions $\phi_1, \phi_2, \ldots, \phi_r$ for (4.12) with $r < n$. Then, the $n \times r$ matrix

$$\Psi(t) \equiv (\phi_1 \, \phi_2 \ldots \phi_r)$$

will have rank r. We will suppose that Ψ can be split as

$$\Psi(t) \equiv \begin{pmatrix} \Psi_1(t) \\ \Psi_2(t) \end{pmatrix},$$

where the $r \times r$ matrix $\Psi_1(t)$ is *nonsingular* for all t of interest (this might require the rows of x to be reordered or the interval of t values to be restricted so that the upper $r \times r$ block of $\Psi(t)$ remains nonsingular). Let us next introduce the nonsingular $n \times n$ matrix

$$T(t) \equiv \begin{pmatrix} \Psi_1(t) & 0 \\ \Psi_2(t) & I \end{pmatrix} \tag{4.13}$$

(using the $(n-r) \times (n-r)$ identity matrix I) and the so-called kinematic change of variables

$$x(t) \equiv T(t)y(t). \tag{4.14}$$

Then, $\dot{x} = \dot{T}y + T\dot{y} = ATy$ implies that y will satisfy

$$\dot{y} = T^{-1}(AT - \dot{T})y = T^{-1}A \begin{pmatrix} 0 & 0 \\ 0 & I \end{pmatrix} y,$$

since

$$\dot{T} = \begin{pmatrix} \dot{\Psi}_1 & 0 \\ \dot{\Psi}_2 & 0 \end{pmatrix} = A \begin{pmatrix} \Psi_1 & 0 \\ \Psi_2 & 0 \end{pmatrix}$$

follows because (4.12) implies that $\dot{\Psi} = A\Psi$. If we now split

$$y = \begin{pmatrix} y_1 \\ y_2 \end{pmatrix} \quad \text{and } A = \begin{pmatrix} A_{11} & A_{12} \\ A_{21} & A_{22} \end{pmatrix}$$

after their first r rows (and columns) and note that

$$T^{-1} = \begin{pmatrix} \Psi_1^{-1} & 0 \\ -\Psi_2\Psi_1^{-1} & I \end{pmatrix},$$

we get the block-triangular system

$$\dot{y}_1 = \Psi_1^{-1}(t)A_{12}(t)y_2$$
$$\dot{y}_2 = [A_{22}(t) - \Psi_2(t)\Psi_1^{-1}(t)A_{12}(t)]y_2$$

for y. Thus, we merely have to integrate the $(n-r)$th-order system for y_2. Knowing y_2, perhaps approximately through a numerical integration, we can get y_1 by ordinary integration of its derivative. This process of solving the decoupled, lower-order system for y_2 is far simpler, computationally, than directly integrating the full nth-order system for x. More thoroughly, if we define Ω to be a nonsingular fundamental matrix solution of the $(n-r)$th-order system

$$\dot{\Omega} = (A_{22} - \Psi_2\Psi_1^{-1}A_{12})\Omega, \tag{4.15}$$

we will have $y_2(t) = \Omega(t)\Omega^{-1}(0)y_2(0)$ and thereby

$$y_1(t) = y_1(0) + \left[\int_0^t \Psi_1^{-1}(s)A_{12}(s)\Omega(s)ds\right]\Omega^{-1}(0)y_2(0).$$

To summarize, we have determined the fundamental matrix

$$\Lambda(t) \equiv \begin{pmatrix} I & \int^t \Psi_1^{-1}(s)A_{12}(s)\Omega(s)ds \\ 0 & \Omega(t) \end{pmatrix}$$

for the y system and it implies a fundamental matrix $\Phi(t) \equiv T(t)\Lambda(t)$ for the original system, namely

$$\Phi(t) \equiv \begin{pmatrix} \Psi_1(t) & \Psi_1(t)\int^t \Psi_1^{-1}(s)A_{12}(s)\Omega(s)ds \\ \Psi_2(t) & \Omega(t) + \Psi_2(t)\int^t \Psi_1^{-1}(s)A_{12}(s)\Omega(s)ds \end{pmatrix}. \tag{4.16}$$

It can now be directly checked that $\dot{\Phi}(t) = A(t)\Phi(t)$ and, less simply, that Φ is nonsingular for all t values for which Ψ_1 remains invertible.

If $X(t)$ is a matrix solution of (4.12) for a continuous state matrix $A(t)$, the determinant $W(t) = \det(X(t))$ can be shown to satisfy the scalar differential equation

$$\dot{W}(t) = \mathrm{tr}(A(t))W(t),$$

where the *trace*, $\mathrm{tr}(A(t))$, is the sum of the diagonal entries of $A(t)$. Since

$$W(t) = e^{\int_{t_0}^t \mathrm{tr}(A_{(s)})ds}W(t_0)$$

generalizes Abel's formula, $X(t)$ will remain nonsingular if and only if its *Wronskian* determinant $W(t)$ is nowhere zero. This motivates the critical search for a local set of linearly independent solutions.

4.2 The matrix exponential

For any constant square matrix A, we will now formally define the matrix exponential e^{At} using the infinite series

$$e^{At} = I + At + \frac{1}{2!}A^2t^2 + \ldots + \frac{1}{n!}A^nt^n + \ldots, \tag{4.17}$$

which can (elementwise) be shown to have an infinite radius of convergence with respect to t. Note that termwise differentiation yields

$$\frac{d}{dt}e^{At} = A + \frac{2}{2!}A^2t + \ldots + \frac{n}{n!}A^nt^{n-1} + \ldots = Ae^{At}.$$

Since $e^{At} = I$ at $t = 0$, e^{At} is, thereby, a fundamental matrix for the autonomous vector system $\dot{x} = Ax$. The uniqueness theorem guarantees that the two definitions of the matrix exponential (i.e., as a nonsingular matrix solution and as the power series) actually coincide. [Taking $t = 1$ in (4.17) also defines the matrix e^A even when the matrix $A(t)$ is not constant, as in example 6. Then, however, $e^{A(t)}$ won't usually satisfy $\dot{X} = A(t)X$.]

We would like to obtain efficient ways of calculating e^{At} for various constant matrices A. For example, we can avoid summing the infinite series of matrices in (4.17) in the following two diagonal cases:

(i) If

$$A = \begin{pmatrix} a_1 & & & \\ & a_2 & & \\ & & \ddots & \\ & & & a_n \end{pmatrix},$$

(with all off-diagonal entries being trivial),

$$A^k = \begin{pmatrix} a_1^k & & & \\ & a_2^k & & \\ & & \ddots & \\ & & & a_n^k \end{pmatrix}$$

for every k, so

$$e^{At} = \begin{pmatrix} e^{a_1t} & & & \\ & e^{a_2t} & & \\ & & \ddots & \\ & & & e^{a_nt} \end{pmatrix}$$

is also diagonal.

(ii) If A can be diagonalized (i.e., if we can factor $A = VDV^{-1}$ for some diagonal matrix D and some nonsingular V), so can e^{At}. This follows because $A^k = VD^kV^{-1}$ implies that

$$e^{At} = VV^{-1} + (VDV^{-1})t + \ldots + \frac{1}{n!}(VDV^{-1})^n t^n + \ldots$$
$$= V\left(I + Dt + \ldots + \frac{1}{n!}D^n t^n + \ldots\right)V^{-1} = Ve^{Dt}V^{-1}.$$

In particular, suppose the $n \times n$ matrix A has n linearly independent eigenvectors $v_1, \ldots, v_n$ defined as nontrivial solutions of $(A - \lambda_j I)v_j = 0$. Then, if we introduce the nonsingular *modal matrix*

$$V \equiv (v_1\ v_2 \ldots v_n),$$

which satisfies $AV = (Av_1\ Av_2 \ldots Av_n) = (\lambda_1 v_1\ \lambda_2 v_2 \ldots \lambda_n v_n) = VD$, where

$$D \equiv \begin{pmatrix} \lambda_1 & & & \\ & \lambda_2 & & \\ & & \ddots & \\ & & & \lambda_n \end{pmatrix}$$

is diagonal, we obtain the corresponding decompositions

$$A = VDV^{-1} \quad \text{and} \quad e^{At} = Ve^{Dt}V^{-1}. \tag{4.18}$$

Note that (4.18) does not require the λ_ks to be distinct. Indeed, matrix theory implies that this diagonalization is possible for the important broad class of *real symmetric matrices* A. Since all solutions of $\dot{x} = Ax$ have the form $x(t) = e^{At}c$ for some constant vector c, we will then have $x(t) = Ve^{Dt}k$ for the constant vector $k = V^{-1}c$. This result corresponds to our earlier conclusion that solutions of $\dot{x} = Ax$ for a diagonalizable matrix A are linear combinations of the appropriate vectors $v_i e^{\lambda_i t}$.

More generally, if

$$x(0) = \sum_{i=1}^{m} c_i V_i,$$

for an eigenvector V_i of A corresponding to λ_i, the solution of the corresponding initial value problem will be simply given by

$$x(t) = e^{At}x(0) = \sum_{i=1}^{m} C_i e^{At} V_i = \sum_{i=1}^{m} C_i e^{\lambda_i t} V_i.$$

We'll henceforth concentrate on more exceptional situations when $x(0)$ is not so simply represented (i.e., generalized eigenvectors are needed).

Example 8 The matrix

$$A = \begin{pmatrix} 4 & 2 \\ 3 & 3 \end{pmatrix}$$

has eigenpairs

$$\left(1, \begin{pmatrix} 2 \\ -3 \end{pmatrix}\right) \quad \text{and} \quad \left(6, \begin{pmatrix} 1 \\ 1 \end{pmatrix}\right),$$

so using

$$V = \begin{pmatrix} 2 & 1 \\ -3 & 1 \end{pmatrix}, D = \begin{pmatrix} 1 & 0 \\ 0 & 6 \end{pmatrix}, \quad \text{and} \quad V^{-1} = \frac{1}{5}\begin{pmatrix} 1 & -1 \\ 3 & 2 \end{pmatrix},$$

we obtain the decomposition

$$e^{At} = Ve^{Dt}V^{-1} = \frac{1}{5}\begin{pmatrix} 2e^t + 3e^{6t} & -2(e^t - e^{6t}) \\ -3(e^t - e^{6t}) & 3e^t + 2e^{6t} \end{pmatrix}.$$

For the general determination of e^{At} with A constant, let us begin with the unique factorization

$$p(\lambda) \equiv |A - \lambda I| = (-1)^n(\lambda - \lambda_1)^{m_1}(\lambda - \lambda_2)^{m_2} \cdot \ldots \cdot (\lambda - \lambda_r)^{m_r} = 0 \tag{4.19}$$

of the characteristic polynomial, where $\lambda_1, \ldots, \lambda_r$ are the r distinct eigenvalues of A with corresponding multiplicities $m_1, \ldots, m_r$, which sum to n. Using partial fractions, we can always obtain

$$\frac{1}{p(\lambda)} \equiv \frac{a_1(\lambda)}{(\lambda - \lambda_1)^{m_1}} + \frac{a_2(\lambda)}{(\lambda - \lambda_2)^{m_2}} + \ldots + \frac{a_r(\lambda)}{(\lambda - \lambda_r)^{m_r}}, \tag{4.20}$$

where each $a_i(\lambda)$ is a polynomial in λ with degree less than m_i. Let us define $q_i(\lambda) \equiv \frac{p(\lambda)}{(\lambda-\lambda_i)^{m_i}}$, so multiplication of (4.20) by $p(\lambda)$ yields the "decomposition of unity"

$$1 = a_1(\lambda)q_1(\lambda) + a_2(\lambda)q_2(\lambda) + \ldots + a_r(\lambda)q_r(\lambda).$$

Since $p(A) = 0$ by the Cayley-Hamilton theorem, the analogous algebra conveniently provides us with the *spectral decomposition*

$$I = a_1(A)q_1(A) + a_2(A)q_2(A) + \ldots + a_r(A)q_r(A) \equiv \sum_{i=1}^{r} P_i \tag{4.21}$$

of the identity matrix I, with $P_i \equiv a_i(A)q_i(A)$ implying a corresponding decomposition of all n dimensional vectors. We will illustrate the basic result

(4.21) by the following two examples, which further suggest the algebraic significance of such a decomposition.

Example 9 Let A be a 2×2 matrix with distinct eigenvalues λ_i, so the characteristic polynomial has the factorization $p(\lambda) = |A - \lambda I| = (\lambda - \lambda_1)(\lambda - \lambda_2)$ for $\lambda_1 \neq \lambda_2$. The identities

$$\frac{1}{p(\lambda)} = \frac{1}{\lambda_1 - \lambda_2}\left(\frac{1}{\lambda - \lambda_1} - \frac{1}{\lambda - \lambda_2}\right)$$

and

$$1 = \frac{1}{\lambda_1 - \lambda_2}((\lambda - \lambda_2) - (\lambda - \lambda_1))$$

correspond to (4.20) for $r = 2$ with the common constant $a_1 = -a_2 = (\lambda_1 - \lambda_2)^{-1}$, $q_1(\lambda) = \lambda - \lambda_2$, and $q_2(\lambda) = \lambda - \lambda_1$. The resulting spectral decomposition (4.21) of the identity is simply

$$I = P_1 + P_2,$$

where $P_1 \equiv a_1 q_1(A) \equiv \frac{A - \lambda_2 I}{\lambda_1 - \lambda_2}$, since $q_1(A) = A - \lambda_2 I$, and $P_2 \equiv a_2 q_2(A) \equiv \frac{\lambda_1 I - A}{\lambda_1 - \lambda_2}$, as should be directly verified. Note, further, that $P_j^2 = P_j$ follows from the Cayley-Hamilton theorem for both $j = 1$ and 2 that $P_j v_j = v_j$ for the eigenvector v_j which satisfies $(A - \lambda_j I)v_j = 0$, and that $P_k v_j = 0$ whenever $k \neq j$. Thus, the matrix P_1 projects any two-dimensional vector $v = \alpha v_1 + \beta v_2$ to its component αv_1 in the one-dimensional nullspace of $A - \lambda_1 I$; P_2 projects such vectors onto the orthogonal nullspace of $A - \lambda_2 I$. Exercise 19 further illustrates how to use this decomposition.

Example 10 Suppose A is a 3×3 matrix with only two distinct eigenvalues λ_j such that $p(\lambda) = |A - \lambda I| = -(\lambda - \lambda_1)^2(\lambda - \lambda_2)$ for $\lambda_1 \neq \lambda_2$. Applying partial fractions to $(p(\lambda))^{-1}$ shows that

$$1 = a_1(\lambda)(\lambda - \lambda_2) + a_2(\lambda)(\lambda - \lambda_1)^2,$$

for the linear function $a_1(\lambda) = \frac{\lambda - 2\lambda_1 - \lambda_2}{(\lambda_1 - \lambda_2)^2}$ and the constant function $a_2 = \frac{-1}{(\lambda_1 - \lambda_2)^2}$, as should be directly verified. We likewise obtain the spectral decomposition

$$I = P_1 + P_2$$

of the 3×3 identity matrix using the projections

$$\begin{aligned} P_1 &\equiv [A + (-2\lambda_1 - \lambda_2)I]\frac{(A - \lambda_2 I)}{(\lambda_1 - \lambda_2)^2} \\ &= \frac{1}{(\lambda_1 - \lambda_2)^2}\left[A^2 - 2\lambda_1 A - \lambda_2(\lambda_2 - 2\lambda_1)I\right] \end{aligned}$$

and

$$P_2 = -\left(\frac{A-\lambda_1 I}{\lambda_1-\lambda_2}\right)^2 = \frac{-1}{(\lambda_1-\lambda_2)^2}\left[A^2 - 2\lambda_1 A + \lambda_1^2 I\right].$$

Again, the spectral decomposition of I can be directly verified as can $P_j^2 = P_j$ for $j = 1$ and 2. Moreover, $P_2 v_2 = v_2$ whenever v_2 satisfies $(A_2 - \lambda_2 I)v_2 = 0$, so P_2 projects onto the one-dimensional nullspace of $A - \lambda_2 I$, whereas the matrix P_1 projects onto the corresponding complementary two-dimensional (possibly generalized) nullspace of $A - \lambda_1 I$.

We shall now use the spectral decomposition (4.21) for any constant matrix A to rewrite

$$e^{At} \equiv \left(\sum_{i=1}^{r} P_i\right) e^{At} = \sum_{i=1}^{r} a_i(A) q_i(A) e^{\lambda_i t} e^{(A-\lambda_i I)t}.$$

However,

$$q_i(A)e^{(A-\lambda_i I)t} = q_i(A)\sum_{j=0}^{\infty}(A-\lambda_i I)^j \frac{t^j}{j!} = q_i(A)\sum_{j=0}^{m_i-1}(A-\lambda_i I)^j \frac{t^j}{j!}$$

because the Cayley-Hamilton theorem implies that $q_i(A)(A-\lambda_i I)^j = p(A)(A-\lambda_i I)^{j-m_i} = 0$ for all $j \geq m_i$. Thus, we finally have the convenient explicit finite representation

$$e^{At} = \sum_{i=1}^{r} e^{\lambda_i t} P_i \sum_{j=0}^{m_i-1} (A-\lambda_i I)^j t^j / j! \tag{4.22}$$

of the matrix exponential, which clearly implies that all (vector or matrix) solutions of $\dot{x} = Ax$ are superpositions of scalar exponentials $e^{\lambda_i t}$ times (vector or matrix) polynomials in t of degree less than the multiplicity m_i of the eigenvalue λ_i as a root of the characteristic polynomial. As already shown, this allows us to compute e^{At} by a scheme of undetermined matrix coefficients when we set

$$e^{At} = \sum_{i=1}^{r} e^{\lambda_i t} \sum_{j=0}^{m_i-1} c_{ji} t^j / j! \tag{4.23}$$

and successively determine the matrices c_{ji} termwise. Moreover, (4.21) and (4.22) show that the matrices $P_j \equiv a_j(A)q_j(A)$ project any vector onto its component in the m_j-dimensional subspace of the solution space for $\dot{x} = Ax$ corresponding to the eigenvalue λ_j. Special handling of complex eigenvalues can also provide real solutions, if desired. Because the sums in (4.23) are finite, convergence of e^{At} for all t is not in question.

An alternative representation of the matrix exponential e^A can be obtained by dividing the function e^λ by the characteristic polynomial $p(\lambda)$ of A to get

$$e^\lambda = p(\lambda)g(\lambda) + r(\lambda),$$

where the remainder $r(\lambda)$ is a polynomial with degree less than n. Because $p(\lambda_i) = 0$ for every eigenvalue λ_i of A, the equations

$$e^{\lambda_i} = r(\lambda_i)$$

can be readily used to find the coefficients in the remainder $r(\lambda)$, at least when A has n distinct eigenvalues. More generally, the Cayley-Hamilton theorem can be used to show that

$$e^A = r(A).$$

Example 11 The matrix

$$A = \begin{pmatrix} -5 & -5 & -9 \\ 8 & 9 & 18 \\ -2 & -3 & -7 \end{pmatrix}$$

has the characteristic polynomial

$$p(\lambda) = |A - \lambda I| = \begin{vmatrix} -5-\lambda & -5 & -9 \\ 8 & 9-\lambda & 18 \\ -2 & -3 & -7-\lambda \end{vmatrix} = -(\lambda+1)^3 = 0.$$

Since $p(A) = -(A+I)^3 = 0$, A has only one eigenvalue, $\lambda = -1$, and

$$e^{At} = e^{-t}e^{(A+I)t} = e^{-t}\left[I + (A+I)t + (A+I)^2\frac{t^2}{2}\right]$$

$$= e^{-t}\left[\begin{pmatrix} 1 & 0 & 0 \\ 0 & 1 & 0 \\ 0 & 0 & 1 \end{pmatrix} + \begin{pmatrix} -4 & -5 & -9 \\ 8 & 10 & 18 \\ -2 & -3 & -6 \end{pmatrix} t + \begin{pmatrix} -4 & -5 & -9 \\ 8 & 10 & 18 \\ -2 & -3 & -6 \end{pmatrix}^2 \frac{t^2}{2}\right],$$

so

$$e^{At} = e^{-t}\begin{bmatrix} 1-4t-3t^2 & -5t-\frac{3}{2}t^2 & -9t \\ 8t+6t^2 & 1+10t+3t^2 & 18t \\ -2t-2t^2 & -3t-t^2 & 1-6t \end{bmatrix}.$$

The solution of the system $\dot{x} = Ax$ for this particular matrix A can also be conveniently obtained using the Jordan form decomposition $A = PJP^{-1}$, where we define the generalized modal matrix

$$P \equiv (p_1\ p_2\ p_3)$$

by using the generalized eigenvectors p_i defined by the *eigenvector chain*

$$(A+I)p_1 = 0, \quad (A+I)p_2 = p_1, \quad \text{and} \quad (A+I)p_3 = p_2.$$

(Trial and error could be used to show that no simpler eigenvector structure is possible.) Because $Ap_1 = -p_1$, $Ap_2 = -p_2 + p_1$, and $Ap_3 = -p_3 + p_2$, these three equations are equivalent to writing $AP = PJ$ for the corresponding Jordan form

$$J \equiv \begin{pmatrix} -1 & 1 & 0 \\ 0 & -1 & 1 \\ 0 & 0 & -1 \end{pmatrix},$$

so we have the factorization $A = PJP^{-1}$ with

$$P = \begin{pmatrix} 3 & 2 & 1 \\ -6 & -4 & -3 \\ 2 & 1 & 1 \end{pmatrix} \quad \text{and} \quad P^{-1} = \begin{pmatrix} 1 & 1 & 2 \\ 0 & -1 & -3 \\ -2 & -1 & 0 \end{pmatrix}.$$

(Readers should check the accuracy of all this algebra.) Writing $J = -I + N$, note that

$$N^2 = \begin{pmatrix} 0 & 0 & 1 \\ 0 & 0 & 0 \\ 0 & 0 & 0 \end{pmatrix},$$

whereas $N^k = 0$ for all integers $k \geq 3$. Thus, $e^{Jt} = e^{-It}e^{Nt} = e^{-t}e^{Nt} = e^{-t}(I + Nt + N^2t^2/2)$, so the decomposition $e^{At} = Pe^{Jt}P^{-1}$ implies that

$$e^{At} = e^{-t} \begin{pmatrix} 3 & 2 & 1 \\ -6 & -4 & -3 \\ 2 & 1 & 1 \end{pmatrix} \begin{pmatrix} 1 & t & t^2/2 \\ 0 & 1 & t \\ 0 & 0 & 1 \end{pmatrix} \begin{pmatrix} 1 & 1 & 2 \\ 0 & -1 & -3 \\ -2 & -1 & 0 \end{pmatrix},$$

which agrees with our previous determination. Finding the Jordan form of a matrix A is basic to determining the behavior of solutions to $\dot{x} = Ax$.

In closing, we note that Moler and Van Loan (1978) consider the practical question of the numerical computation of the matrix exponential. In the case of distinct eigenvalues, they write (4.22) as

$$e^{At} = \sum_{j=1}^{n} e^{\lambda_j t} \prod_{\substack{k=1 \\ k\neq j}}^{n} \frac{(A-\lambda_k I)}{\lambda_j - \lambda_k},$$

interpreting the result as a Lagrange interpolation formula. Otherwise, practical, but unavoidable, complications result from the direct dependence of e^{At} on the Jordan form J of A and the possibility of substantial change in J and e^{At} resulting from small changes in the entries of A which dramatically alter its eigenstructure.

Example 12 The real symmetric matrix

$$A = \begin{pmatrix} 1 & -4 & -1 \\ -4 & -2 & 4 \\ -1 & 4 & 1 \end{pmatrix}$$

has the eigenpairs

$$(\lambda_i, v_i) = \left(6, \begin{pmatrix} -1 \\ 1 \\ 1 \end{pmatrix}\right), \left(0, \begin{pmatrix} 1 \\ 0 \\ 1 \end{pmatrix}\right), \quad \text{and} \quad \left(-6, \begin{pmatrix} -1 \\ -2 \\ 1 \end{pmatrix}\right)$$

and satisfies its characteristic polynomial $-A^3 + 36A = 0$. Moreover, any vector v can be written as a linear combination $\alpha_1 v_1 + \alpha_2 v_2 + \alpha_3 v_3$ of the basis vectors v_i. Reflecting the partial fraction expansion

$$\frac{1}{-\lambda^3 + 36\lambda} = \frac{1}{72}\frac{1}{(\lambda-6)} - \frac{1}{36\lambda} + \frac{1}{72(\lambda+6)},$$

and the decomposition

$$1 = \frac{\lambda}{72}(\lambda+6) - \frac{1}{36}(\lambda^2 - 36) + \frac{\lambda}{72}(\lambda-6),$$

the projection matrices

$$P_1 \equiv \frac{1}{72}A(A+6I) = \frac{1}{72}A^2 + \frac{1}{12}A = \frac{1}{3}\begin{bmatrix} 1 & -1 & -1 \\ -1 & 1 & 1 \\ -1 & 1 & 1 \end{bmatrix},$$

$$P_2 \equiv -\frac{1}{36}(A-6I)(A+6I) = -\frac{1}{36}A^2 + I = \frac{1}{2}\begin{bmatrix} 1 & 0 & 1 \\ 0 & 0 & 0 \\ 1 & 0 & 1 \end{bmatrix},$$

and

$$P_3 \equiv \frac{1}{72}(A - 6I)A = \frac{1}{72}A^2 - \frac{1}{12}A = \frac{1}{6}\begin{bmatrix} 1 & 2 & -1 \\ 2 & 4 & -2 \\ -1 & -2 & 1 \end{bmatrix}$$

satisfy $P_1 + P_2 + P_3 = I$ and $P_j v_j = v_j$ for $j = 1, 2$, and 3. Since the eigenvalues of A are all simple, (4.22) implies the symmetric matrix exponential

$$\begin{aligned} e^{At} &= e^{6t}P_1 + P_2 + e^{-6t}P_3 \\ &= \frac{1}{6}\begin{bmatrix} 2e^{6t} + 3 + e^{-6t} & 2(-6e^{6t} + e^{-6t}) & -2e^{6t} + 3 - e^{-6t} \\ 2(-6e^{-6t} + e^{-6t}) & 2e^{6t} + 4e^{-6t} & 2(e^{6t} - e^{-6t}) \\ -2e^{6t} + 3 - e^{-6t} & 2(e^{6t} - e^{-6t}) & 2e^{6t} + 3 + e^{-6t} \end{bmatrix}, \end{aligned}$$

as can be directly checked.

The treatment of constant systems, in turn, allows us to construct series solutions of certain systems $\dot{x} = A(t)x$. In the regular case that $A(t)$ has a Maclaurin expansion, we can expect to obtain the Maclaurin expansion for a corresponding fundamental matrix. If we likewise consider

$$t\dot{X} = A(t)X,$$

where A has a Maclaurin expansion, we would naturally call $t = 0$ a regular singular point and thereby seek a fundamental matrix in the Frobenius form

$$X(t) = t^D P(t),$$

where D is a constant matrix and the matrix $P(t)$ has its own Maclaurin expansion. Certainly, $t^D \equiv e^{D \ln t}$ must first be determined and we must expect breakdowns (i.e., the appearance of logarithms in the solution) when the eigenvalues of $A(0)$ are repeated or differ by integers. These ideas will be explored, to a limited extent, in the exercises.

4.3 Exercises

1 Solve $\dot{x} = Ax$ for the real symmetric matrix

$$A = \begin{pmatrix} 1 & -4 & -1 \\ -4 & -2 & 4 \\ -1 & 4 & 1 \end{pmatrix}.$$

2 For what initial values will the system $\dot{x} = Ax$ for $A = \begin{pmatrix} -1 & 6 \\ 1 & -2 \end{pmatrix}$ have bounded solutions for all $t \geq 0$?

3 Solve the equation

$$\dot{x} = \begin{bmatrix} B & C(t) \\ 0 & D \end{bmatrix} x + g(t)$$

in terms of e^{Bt} and e^{Dt} when x and g are n-vectors, B is a constant $k \times k$ matrix, and D is a constant $(n-k) \times (n-k)$ matrix.

4 (a) If A and B are constant $n \times n$ matrices that satisfy $AB = BA$, show that $e^{A+B} = e^A e^B$.

(b) Use (a) to show that the matrix e^A has the inverse e^{-A}.

5 (a) Show that the second-order matrix differential equation

$$Y'' + A^2 Y = 0,$$

for a constant square matrix A, is satisfied by

$$\cos At \equiv I - \frac{1}{2}A^2t^2 + \frac{1}{24}A^4t^4 - \dots .$$

(b) Find the closed-form general solution for the vector equation $y'' + N^2 y = 0$ with $N = \begin{bmatrix} 0 & 1 & 0 \\ 0 & 0 & 1 \\ 0 & 0 & 0 \end{bmatrix}$.

6 Consider the singular linear matrix system

$$xY' = A(x)Y,$$

where $A(x) = \sum_{r=0}^{\infty} A_r x^r$. Suppose we make the change of variables

$$Y = P(x)Z,$$

where Z is a fundamental matrix for the simplified (limiting) differential system

$$xZ' = A_0 Z.$$

What singular linear system must the square matrix $P(x)$ satisfy?

7 Use the series expansion for e^{Jt} to find a closed-form solution of $\dot{x} = Jx$ when

$$J = \begin{pmatrix} 0 & -1 \\ 1 & 0 \end{pmatrix}.$$

(Note that $J^2 = -I$.)

8 Suppose the matrix $Y(x) = \exp[\int_0^x A(s)ds]$ is defined using the exponential series. Presuming $Y(x)$ is a nonsingular solution of the second-order linear matrix equation

$$Y'' + B(x)Y' + C(x)Y = 0$$

(and that Y commutes with A, A^2, A' and $\int_0^x A(s)ds$), show that $A(x)$ must satisfy a quadratic matrix equation.

9 Determine all possible solutions of the linear *differential-algebraic* equation defined by the system $\dot{x} = Ax$ and the constraint $Cx = 0$ when x is an n-vector and C is an $m \times n$ matrix of rank $m < n$.

10 Suppose we seek to move the *state* $x(t)$ of a system from a prescribed initial vector x_0 at $t = t_0$ to a prescribed terminal value x_1 at $t = t_1$ by letting a *control vector* $u(t)$ drive the linear system

$$\dot{x}(t) = A(t)x(t) + B(t)u(t).$$

Let $\Phi(t, t_0)$ be the fundamental matrix for the unforced system $\dot{\Phi} = A(t)\Phi$ with $\Phi(t_0) = I$, and introduce the *control law* $u(t) = -B'(t)\Phi'(t, t_0)\eta$ for some unknown constant vector η (where the primes denote transposition). Show that solutions $x(t)$ exist whenever the vector $x_0 - \Phi^{-1}(t_0, t_1)x_1$ lies in the range of the *controllability Gramian* (matrix)

$$G(t_0, t_1) \equiv \int_{t_0}^{t_1} \Phi^{-1}(s, t_0)B(s)B'(s)\Phi'(s, t_0)ds.$$

11 (a) Show that the unique solution of the matrix initial value problem

$$\frac{dZ}{dt} = AZ + ZB, \quad Z(0) = C$$

is $Z(t) = e^{At}Ce^{Bt}$.

(b) By integrating

$$\frac{dZ}{dt} = AZ + ZB$$

from $t = 0$, where $Z(0) = C$, to $t = \infty$, where $Z(\infty) = 0$, show that the *Liapunov equation*

$$AX + XB = C$$

has the solution $X = -\int_0^\infty e^{As}Ce^{Bs}ds$, presuming this integral exists.

12 Suppose X is a nonsingular solution of $X' = AX$ for a constant matrix A. Differentiate $XX^{-1} = I$ to obtain the linear matrix equation satisfied by X^{-1}.

13 (a) Beginning with

$$f(x) = f(x_0) + \int_{x_0}^{x} f'(t)dt,$$

use repeated integration by parts to show that

$$f(x) = f(x_0) + f'(x_0)(x - x_0) + \ldots + \frac{f^{(n)}}{n!}(x_0)(x - x_0)^n + \int_{x_0}^{x} \frac{(x-t)^n}{n!} f^{(n+1)}(t)dt$$

on an interval about x_0 where f is sufficiently smooth.

(b) Show that the difference between a given smooth function $f(x)$ and its nth Taylor polynomial

$$P_n(x) \equiv \sum_{j=0}^{n} \frac{f^{(j)}(x_0)}{j!}(x - x_0)^j$$

satisfies

$$|f(x) - P_n(x)| \leq \frac{M_n}{(n+1)!}|x - x_0|^{n+1}$$

on $|x - x_0| \leq R$, where $M_n = \max_{|x-x_0|\leq R} |f^{(n+1)}(x)|$.

(c) Using the periodicity of $\sin x$, devise a way to approximate $\sin x$ for all x values by using its Taylor polynomials about $x = 0$ for $0 \leq x \leq \frac{\pi}{2}$.

14 Consider the system $\dot{x} = Ax$ for $A = \begin{pmatrix} 1 & -5 \\ 1 & -1 \end{pmatrix}$.

(a) Seek a vector solution $x = e^{\lambda t} v$ where λ is an eigenvalue of A.

(b) Define the matrix solution $\Phi(t) = (e^{\lambda t} v \;\; e^{\overline{\lambda} t}\overline{v})$ and compare $\Phi(t)\Phi^{-1}(0)$ with e^{At}.

15 (a) Solve the vector initial value problem

$$x' = Ax + be^t, t \geq 0$$

when the eigenvalues λ of A satisfy $Re\lambda < 1$.

(b) Determine the *sensitivity* $\frac{dx(t)}{dx(0)}$ of the solution of the preceding initial value problem to changes in the initial vector $x(0)$.

(c) Show that the steady state (i.e., long-time behavior) is independent of $x(0)$.

(d) Show that (c) no longer holds when $(A - I)^{-1}$ exists, but $(A - 2I)^{-1}$ does not.

16 Show that the linear system

$$x' = A(t)x + u$$

with the variable control

$$u(t) = -A(t)x(0)$$

will maintain a constant state $x(t)$.

17 (a) Given

$$A = \begin{bmatrix} 1 & 0 \\ 0 & 0 \end{bmatrix} \quad \text{and } B = \begin{bmatrix} 0 & 1 \\ 0 & 0 \end{bmatrix},$$

compare AB and BA.

(b) Compute e^{At} and e^{Bt} and compare $e^{(A+B)t}$, $e^{At}e^{Bt}$, and $e^{Bt}e^{At}$.

18 Use the formula $\cos\theta = \frac{1}{2}(e^{i\theta} + e^{-i\theta})$ to determine the matrix

$$\cosh At = \cos(iAt) \quad \text{for } A = \begin{bmatrix} -1 & 1 & 0 \\ -1 & -1 & 0 \\ 0 & 0 & 2 \end{bmatrix}.$$

19 (a) Suppose A is a real 2×2 matrix with complex eigenvalues $\lambda = \alpha \pm i\beta$. Use the spectral decomposition

$$I = \frac{1}{\lambda - \bar{\lambda}}[(A - \bar{\lambda}I) - (A - \lambda I)]$$

and the representation (4.22) to obtain

$$e^{At} = e^{\alpha t}\left[\frac{1}{\beta}(\beta\cos\beta t - \alpha\sin\beta t)I + \sin\beta t A\right].$$

(b) Carry out the calculation for $A = \begin{pmatrix} 1 & -5 \\ 1 & -1 \end{pmatrix}$.

20 (a) For

$$A = \begin{pmatrix} 1 & 0 & 0 \\ 4 & 1 & 0 \\ 5 & 3 & 2 \end{pmatrix},$$

show that $A^2 - 2A$ is a projection onto the two-dimensional nullspace of $A - I$.

(b) Use (4.22) to show that

$$e^{At} = e^t(A^2 - 2A)[I + (A - I)t] - e^{2t}[A^2 - 2A + I].$$

(c) Verify that the matrix e^{At} obtained in (b) satisfies $\dot{x} = Ax$.

21 Find a matrix function f that satisfies

$$f(s+t) = f(s)f(t)$$

for all real values s and t among all continuously differentiable matrix functions f such that $f(0) = I$ and $f'(0) = A$. [Hint: Find $f'(s) = \lim_{h\to 0}\left(\frac{f(s+h)-f(s)}{h}\right)$.]

22 Suppose a matrix $X(t)$ satisfies the initial value problem

$$\frac{dx}{dt} = (A + \epsilon B(t))x, \quad x(0) = I$$

for a constant matrix A, a small parameter ϵ, and a continuous matrix $B(t)$.

(a) Convert the initial value problem for x to an integral equation for $z = e^{-At}x$. Then, iterate in the integral equation to find the first few terms in a series expansion for z in powers of ϵ.

(b) Show that the results obtained in (a) agree with the alternative of using a formal power series

$$X(t) = X_0(t) + \epsilon X_1(t) + \epsilon^2 X_2(t) + \ldots$$

in the given initial value problem (at least for bounded t values).

23 Consider the linear nonhomogeneous vector system

$$\dot{x} = Ax + e^{\alpha t}b,$$

where A, α, and b are given. Seek a particular solution in the form

$$x(t) = e^{\alpha t}c$$

for some vector c, presuming α is not an eigenvalue of the matrix A. What complication occurs when α is such an eigenvalue? What happens when α approaches such an eigenvalue?

24 (a) Let $x(t)$ and $y(t)$ satisfy the differential system

$$\dot{x} = A(t)x + f(t)$$

with the respective initial vectors $x(0)$ and $y(0)$. Determine a bound for $|x(t) - y(t)|$ in terms of a fundamental matrix $\Phi(t)$ for $\dot{z} = A(t)z$ and $|x(0) - y(0)|$.

(b) If *all* solutions of the homogeneous system $\dot{z} = A(t)z$ decay to zero as $t \to \infty$, show that $x(t) \to y(t)$ as $t \to \infty$.

25 (a) If A is a nonsingular matrix, show that

$$\int^t e^{At}\,dt = A^{-1}e^{At}$$

(b) If A is singular, use the representations (22) and (23) to determine $\int^t e^{As}\,ds$.

26 Suppose two connected water tanks A and B each contain 100 gallons of water. Water flows into A at 4 gallons/minute and out of B at the same rate. Suppose 6 gallons/minute flows from A to B, while 4 gallons/minute flows from B to A. If a pound of a toxic substance is dropped into tank B at $t = 0$, find the amount of the substance in both tanks for all later times t, assuming the inflow remains pure, the tanks are kept stirred, and there is no further contamination. (The basic approach suggested here relates to *compartmental models*, which are extensively used in economic modeling and in the biosciences.)

27 (a) Consider the system

$$t\dot{X} = (A_0 + tI)X \text{ for } t > 0,$$

where $A_0 + tI = \begin{pmatrix} \frac{5}{6} + t & -1 \\ \frac{2}{3} & -1 + t \end{pmatrix}$. Since A_0 has eigenpairs $\left(\frac{1}{3}, \begin{pmatrix} 2 \\ 1 \end{pmatrix}\right)$ and $\left(-\frac{1}{2}, \begin{pmatrix} 3 \\ 4 \end{pmatrix}\right)$, seek solutions of the system in the form

$$\varphi_1(t) = t^{1/3}\begin{pmatrix} 2 \\ 1 \end{pmatrix} p_1(t)$$

and

$$\varphi_2(t) = t^{-1/2}\begin{pmatrix} 3 \\ 4 \end{pmatrix} p_2(t)$$

for scalar functions p_1 and p_2.

(b) Motivated by the eigenfunction analysis of (a), show that one can analogously find a solution of the system

$$t\dot{X} = \begin{pmatrix} 1 & t \\ 0 & 0 \end{pmatrix} X \text{ for } t > 0$$

in the form $X_1(t) = t\begin{pmatrix} 1 \\ 0 \end{pmatrix} p_1(t)$, but not as $X_2(t) = \begin{pmatrix} 0 \\ 1 \end{pmatrix} p_2(t)$. Verify that a linearly independent solution is given by $\tilde{X}_2(t) \equiv \begin{pmatrix} t \ln t \\ 1 \end{pmatrix}$.

28 (a) Find a power series solution of the vector system

$$t^2\dot{X} = (A_0 + A_1 t)X, \quad X(0) = \begin{pmatrix} 1 \\ 0 \end{pmatrix},$$

where $A_0 + A_1 t = \begin{pmatrix} 0 & t \\ -t & -1-2t \end{pmatrix}$. (Hint: The first component of X satisfies the equation of example 11 in Chapter 3.)

(b) Where does the series converge?

29 Solve the matrix initial value problem

$$t^3\dot{X} = AX, \quad X(1) = I$$

for a constant matrix A and $t > 0$. (Hint: Let the scalar problem be your guide.)

30 Consider the complex-valued linear system

$$\dot{X} = A(t)X,$$

where $A(t) \equiv B(t) + iC(t)$ for real matrices B and C.

(a) Let $X(t) \equiv U(t) + iV(t)$ for real vectors U and V. Show that

$$W(t) \equiv \begin{pmatrix} U(t) \\ V(t) \end{pmatrix}$$

satisfies the real-valued linear system

$$\dot{W} = \mathcal{A}(t)W$$

for $\mathcal{A}(t) \equiv \begin{pmatrix} B & -C \\ C & B \end{pmatrix}$.

(b) If $X \equiv U + iV$ is a fundamental matrix for the complex-valued system, show that

$$W = \begin{pmatrix} U & -V \\ V & U \end{pmatrix}$$

will be a fundamental matrix for the real-valued system. [Hint: To show the nonsingularity of W, note that

$$PWP^{-1} = \begin{pmatrix} X & 0 \\ 0 & \overline{X} \end{pmatrix} \text{ for } P = \begin{pmatrix} I & -iI \\ -iI & I \end{pmatrix}.]$$

31 Show that all solutions of the linear matrix equation

$$y' = Ay + yB + G(t)$$

for constant matrices A and B are given by

$$y(t) = e^{At}y(0)e^{Bt} + \int_0^t e^{A(t-s)}G(s)e^{B(t-s)}ds.$$

32 Show that the *Prüfer substitution* $u = \rho\cos\theta$, $v = \rho\sin\theta$ converts the two-dimensional system

$$\begin{cases} u' = a_{11}(x)u + a_{12}(x)v \\ v' = a_{21}(x)u + a_{22}(x)v \end{cases}$$

to the form

$$\begin{cases} \theta' = \frac{1}{2}(a_{21} - a_{12}) + \frac{r}{2}\cos(2\theta + \psi) \\ \frac{\rho'}{\rho} = \frac{1}{2}(a_{11} + a_{22}) + \frac{r}{2}\sin(2\theta + \psi), \end{cases}$$

where $r^2 = (a_{11} - a_{22})^2 + (a_{12} + a_{21})^2$, $\cos\psi = \frac{1}{r}(a_{12} + a_{22})$, and $\sin\psi = \frac{1}{r}(a_{11} - a_{22})$. Note that the differential equation for θ is independent of ρ, whereas the equation for ρ is readily solved as a separable equation if we know $\theta(x)$.

33 (a) Consider the nonhomogeneous linear system

$$X' = AX + B(t),$$

where A is a nonsingular constant matrix and $B(t)$ is the quadratic vector

$$B(t) = b_0 + b_1 t + b_2 t^2.$$

Determine a quadratic vector solution

$$X(t) = c_0 + c_1 t + c_2 t^2.$$

(b) Solve

$$X' = AX + \sum_{k=0}^{M} b_k t^k$$

for any finite integer M, assuming A is nonsingular.

(c) Solve

$$X' = AX + e^{\gamma t}(b_0 + b_1 t + b_2 t^2),$$

assuming that γ is not an eigenvalue of A.

(d) If zero is a simple eigenvalue of A (with a corresponding nullvector), find a solution of (a) which is a cubic polynomial in t.

34 Suppose y satisfies the scalar initial value problem

$$y' = f(x, y), \quad y(x_0) = y_0$$

for some continuously differentiable function f. Show

(a) $\frac{\partial y}{\partial y_0} = e^{\int_{x_o}^{x} f_y(t, y(t))dt}$

(b) $\frac{\partial y}{\partial x_0} = -f(x_0, y_0)\frac{\partial y}{\partial y_0}$.

35 (a) Show that the time-varying system

$$\dot{x} = A(t)x,$$

with $A(t) = \begin{bmatrix} -1 + \gamma \cos^2 t & 1 - \gamma \sin t \cos t \\ -1 - \gamma \sin t \cos t & -1 + \gamma \sin^2 t \end{bmatrix}$ and a constant γ, can be solved directly using the orthogonal transformation

$$y = P(t)x \quad \text{for} \quad P(t) = \begin{bmatrix} \cos t & -\sin t \\ \sin t & \cos t \end{bmatrix}.$$

(b) Show that all solutions of $\dot{x} = A(t)x$ decay to zero as $t \to \infty$ when $\gamma < 1$, even though the constant eigenvalues of $A(t)$ both have negative real parts for $\gamma < 2$ (cf. Mattheij and Molenaar (1996)).

36 Consider the linear vector system

$$B(t)\frac{dx}{dt} = C(t)x + k(t),$$

where the matrices $B(t)$ and $C(t)$ and the vector $k(t)$ have continuous entries and

$$B(t) = \begin{bmatrix} M(t) & 0 \\ 0 & 0 \end{bmatrix} \text{ and } C(t) = \begin{bmatrix} N(t) & P(t) \\ Q(t) & R(t) \end{bmatrix}$$

are partitioned in the same way, with both the $r \times r$ matrix $M(t)$ and the $(n - r) \times (n - r)$ matrix $R(t)$ being nonsingular.

(a) Show how you can solve the system by partitioning the vector x after its first r rows and solving the resulting linear equations for the two component vectors.

(b) What initial values must be prescribed at $t = 0$ to determine a unique solution?

37 Consider the linear system

$$\dot{z} = A(t)z \text{ for } A(t) = \begin{pmatrix} 0 & 3e^{-2t} \\ e^{2t} & 0 \end{pmatrix}.$$

(a) Let $u = \begin{pmatrix} x \\ y \end{pmatrix}$ and show that x satisfies the scalar equation $\ddot{x} + 2\dot{x} - 3 = 0$.

(b) Find the general solution for x, and use it to determine y and a fundamental matrix for z.

38 Find a fundamental matrix for the decoupled system

$$xy' = A(x)y$$

for $A(x) = \begin{pmatrix} 2 & x^2 \\ 0 & 1 \end{pmatrix}$ and $x \neq 0$.

39 Write an essay on the topic "What good is the matrix exponential?"

Solutions to exercises 1–11

1 For

$$A = \begin{pmatrix} 1 & -4 & -1 \\ -4 & -2 & 4 \\ -1 & 4 & 1 \end{pmatrix},$$

the characteristic polynomial $\det(A - \lambda I) = -\lambda^3 + 36\lambda = 0$ provides the distinct eigenvalues $\lambda_{1,2,3} = 6, 0$, and -6. The corresponding eigenvectors (nonuniquely) satisfy $(A - \lambda I)p = 0$ and $P \equiv (p_1\; p_2\; p_3)$ defines the modal matrix

$$P = \begin{pmatrix} -1 & 1 & -1 \\ 1 & 0 & -2 \\ 1 & 1 & 1 \end{pmatrix}$$

with the inverse

$$P^{-1} = \frac{1}{6}\begin{pmatrix} -2 & 2 & 2 \\ 3 & 0 & 3 \\ -1 & -2 & 1 \end{pmatrix}.$$

Decomposing

$$A = PDP^{-1} \text{ for } D = \begin{pmatrix} 6 & 0 & 0 \\ 0 & 0 & 0 \\ 0 & 0 & -6 \end{pmatrix}$$

determines the exponential e^{At} found in example 12.

2 The matrix $A = \begin{pmatrix} -1 & 6 \\ 1 & -2 \end{pmatrix}$ has the characteristic polynomial $p(\lambda) = \lambda^2 + 3\lambda - 4 = 0$. Two eigenpairs are $\left(-4, \begin{pmatrix} 2 \\ -1 \end{pmatrix}\right)$ and $\left(1, \begin{pmatrix} 3 \\ 1 \end{pmatrix}\right)$, so the general solution of $\dot{x} = Ax$ is $x(t) = \begin{pmatrix} 2 \\ -1 \end{pmatrix} e^{-4t} c_1 + \begin{pmatrix} 3 \\ 1 \end{pmatrix} e^t c_2$. Boundedness as $t \to \infty$ requires $c_2 = 0$, so we must have $x(0) = \begin{pmatrix} 2 \\ -1 \end{pmatrix} c_1$ for a scalar c_1.

3 If we decompose the system $\dot{x} = \begin{bmatrix} B & C(t) \\ 0 & D \end{bmatrix} x + g(t)$ into the subsystems $\dot{x}_1 = Bx_1 + C(t)x_2 + g_1(t)$ and $\dot{x}_2 = Dx_2 + g_2(t)$, variation of parameters implies that $x_1(t) = e^{Bt}x_1(0) + \int_0^t e^{B(t-s)}[C(s)x_2(s) + g_1(s)]ds$ and $x_2(t) = e^{Dt}x_2(0) + \int_0^t e^{D(t-p)}g_2(p)dp$. Substituting the second result into the first completely determines the solution whatever the initial vectors $x_1(0)$ and $x_2(0)$.

4 (a) By definition, $e^{A+B} = \sum_{j=0}^{\infty} \frac{1}{j!}(A+B)^j$, where $(A+B)^j = (A+B)(A+B)\ldots(A+B) = \sum_{k=0}^{j} {}_jC_k A^k B^{j-k}$ and the scalar ${}_jC_k$ is the binomial coefficient, as long as the order of multiplication is irrelevant (i.e., $AB = BA$). Then, direct multiplication of the two series for e^A and e^B also yields the series for e^{A+B}.

(b) Taking $B = -A$, we get $A + B = 0I$ and $e^{0I} = I$. Thus, $e^{A-A} = e^A e^{-A} = I$ implies that $(e^A)^{-1} = e^{-A}$.

5 (a) If $Y = \cos At \equiv I - \frac{1}{2!}A^2t^2 + \frac{1}{4!}A^4t^4 + \ldots$, termwise differentiation implies that $\frac{d^2Y}{dt^2} = -A^2Y$ and likewise for $Y = A^{-1}\sin At \equiv It - \frac{1}{3!}A^2t^3 + \frac{1}{5!}A^4t^5 - \ldots$ (with the series defined even if A is singular). It follows that the general solution of the vector equation $y'' + A^2y = 0$ is $y(t) = (\cos At)y(0) + (A^{-1}\sin At)y'(0)$, as can be directly verified.

(b) For

$$N = \begin{pmatrix} 0 & 1 & 0 \\ 0 & 0 & 1 \\ 0 & 0 & 0 \end{pmatrix}, \; N^2 = \begin{pmatrix} 0 & 0 & 1 \\ 0 & 0 & 0 \\ 0 & 0 & 0 \end{pmatrix}$$

and $N^j \equiv 0$ for each integer $j > 2$. Thus,

$$\cos Nt \equiv I - \frac{1}{2}N^2t^2 = \begin{pmatrix} 1 & 0 & -t^2/2 \\ 0 & 1 & 0 \\ 0 & 0 & 1 \end{pmatrix}$$

and

$$N^{-1}\sin Nt = It - \frac{1}{6}N^2t^3 = \begin{pmatrix} t & 0 & -\frac{1}{6}t^3 \\ 0 & t & 0 \\ 0 & 0 & t \end{pmatrix},$$

so the general solution of $y'' + N^2y = 0$ is

$$y(t) = \begin{pmatrix} 1 & 0 & -t^2/2 \\ 0 & 1 & 0 \\ 0 & 0 & 1 \end{pmatrix} y(0) + \begin{pmatrix} t & 0 & -t^3/6 \\ 0 & t & 0 \\ 0 & 0 & t \end{pmatrix} y'(0),$$

which also has the polynomial representation

$$y(t) = y(0) + y'(0)t - \frac{1}{2}\begin{pmatrix} 0 \\ 0 \\ y_3(0) \end{pmatrix} t^2 - \frac{1}{6}\begin{pmatrix} 0 \\ 0 \\ y_3'(0) \end{pmatrix} t^3$$

$$= \begin{bmatrix} y_1(0) + y_1'(0)t \\ y_2(0) + y_2'(0)t \\ y_3(0)\left(1 - \frac{t^2}{2}\right) + y_3'(0)\left(t - \frac{t^3}{6}\right) \end{bmatrix}$$

when $y = \begin{pmatrix} y_1 \\ y_2 \\ y_3 \end{pmatrix}$. This result could also be found using a straightforward undetermined coefficients procedure.

6 If $Y = P(x)Z$ satisfies $xY' = A(x)Y$ and $xZ' = A_0Z$, $xY' = x(PZ' + P'Z) = A(x)PZ$ implies that $xPA_0Z + xP'Z = A(x)PZ$, so since Z is nonsingular, P must satisfy the linear matrix equation $xP' = A(x)P - PA_0$. A Maclaurin expansion for P would then be natural, because the singular behavior of Y is already accounted for by Z.

7 If

$$J = \begin{pmatrix} 0 & -1 \\ 1 & 0 \end{pmatrix},\ J^2 = -\begin{pmatrix} 1 & 0 \\ 0 & 1 \end{pmatrix} = -I,$$

so

$$e^{Jt} = \sum_{k=0}^{\infty} \frac{J^k t^k}{k!} = I + Jt - I\frac{t^2}{2!} - J\frac{t^3}{3!} + I\frac{t^4}{4!} + J\frac{t^5}{5!} - \dots$$
$$= I(1 - \frac{t^2}{2!} + \frac{t^4}{4!} - \dots) + J(t - \frac{t^3}{3!} + \frac{t^5}{5!} - \dots)$$
$$= \begin{pmatrix} 1 & 0 \\ 0 & 1 \end{pmatrix} \cos t + \begin{pmatrix} 0 & -1 \\ 1 & 0 \end{pmatrix} \sin t = \begin{pmatrix} \cos t & -\sin t \\ \sin t & \cos t \end{pmatrix}.$$

Thus, solutions of $\dot{x} = Jx$ are given by $e^{Jt}x(0)$. An analogous procedure applies if J instead has block matrix entries 0 and I.

8 If we define $Y(x) = \exp[\int_0^x A(s)ds] \equiv I + \int_0^x A(s)ds + \frac{1}{2!}[\int_0^x A(s)ds]^2 + \dots$, $Y'(x) = Y(x)A(x) = A(x)Y(x)$ and $Y''(x) = A(x)Y'(x) + A'(x)\,Y(x) = [A^2(x) + A'(x)]Y(x)$ and if all matrices involved commute, then $Y'' + B(x)Y' + C(x)Y = 0$ implies that $(A^2 + A' + BA + C)Y = 0$. Because Y is nonsingular, A must satisfy the first-order quadratic matrix equation $A' + BA + A^2 + C = 0$.

9 Since all solutions of the linear system $\dot{x} = Ax$ are of the form $x(t) = e^{At}x(0)$, we will need $Ce^{At}x(0) = 0$ for all t. In particular, $Cx(0) = 0$ limits the n-vector $x(0)$ to having $n - m$ degrees of freedom. Keeping $x(t)$ in the nullspace of C is then required for $t \neq 0$. If, for example, $C = [I\ 0]$, the first m components of x must remain zero. If we write $x = \begin{pmatrix} x_1 \\ x_2 \end{pmatrix}$ and $e^{At} \equiv \begin{pmatrix} \phi_{11}(t) & \phi_{12}(t) \\ \phi_{21}(t) & \phi_{22}(t) \end{pmatrix}$, we will then need to select $x_2(0)$ so that $x_1(t) \equiv \phi_{12}(t)x_2(0) \equiv 0$ as t varies. For a more general matrix C, a preliminary transformation of variables would allow us to assume that C is in the special form.

10 If we set $x(t) = \Phi(t, t_0)z(t)$, as in variation of parameters, we would need $\dot{z} = \Phi^{-1}Bu$, $z(t_0) = x_0$, and $z(t_1) = \Phi^{-1}(t_1, t_0)x_1$, so integration gives $z(t) = x_0 + \int_{t_0}^t \Phi^{-1}(s, t_0)B(s)u(s)ds$. The terminal condition for z therefore requires that $\Phi^{-1}(t_1, t_0)x_1 = z(t_1)$, so invoking the control law requires the vector η to satisfy the linear algebraic system

$$G(t_0, t_1)\eta = \left[\int_{t_0}^{t_1} \Phi^{-1}(s, t_0)B(s)B'(s)\Phi'(s, t_0)ds\right]\eta = x_0 - \Phi^{-1}(t_1, t_0)x_1$$

Any solution η of this system determines a satisfactory control $u(t)$. To get a solution, however, the right-hand side must lie in the range of the Gramian matrix G.

11 (a) Note that $Z(t) = e^{At}Ce^{Bt}$ implies that $\frac{dZ}{dt} = Ae^{At}Ce^{Bt} + e^{At}Ce^{Bt}B$ and that $Z(0) = C$. The conclusion then follows by the uniqueness of solutions to linear constant-coefficient initial value problems. Alternatively, note that the equation is equivalent to $\frac{d}{dt}(e^{-At}Ze^{-Bt}) = 0$, so $e^{-At}Ze^{-Bt} = C$ for the constant matrix $C = Z(0)$.

(b) Integrating the differential equation $\dot{Z} = AZ + ZB$ from 0 to ∞, we have $Z(0) - Z(\infty) = A\int_0^\infty Z(s)ds + \int_0^\infty Z(s)ds\,B$, so (a) implies that $-\int_0^\infty e^{As}Ce^{Bs}ds$ satisfies the given Liapunov equation, presuming the integral exists. Some further effort (i.e., matrix analysis) would show that this integration is possible for all matrices C as long as A and $-B$ have no eigenvalues in common.

5
Stability concepts

5.1 Two-dimensional linear systems

A constant real coefficient system of two scalar linear homogeneous equations has the form

$$\begin{cases} \dot{x} = ax + by \\ \dot{y} = cx + dy \end{cases} \tag{5.1}$$

(for all values of the independent variable t) or, equivalently, the vector-matrix form

$$\dot{z} = Az, \tag{5.2}$$

where

$$z \equiv \begin{pmatrix} x \\ y \end{pmatrix} \text{ and } A \equiv \begin{pmatrix} a & b \\ c & d \end{pmatrix}$$

is a constant matrix with the characteristic polynomial $\det(A - \lambda I) = \lambda^2 - (a + d)\lambda + ad - bc = 0$. Let us first suppose that A has eigenpairs (λ_1, p_1) and (λ_2, p_2) with p_1 and p_2 being linearly independent eigenvectors. Since

$$Ap_i = \lambda_i p_i$$

for $i = 1$ and 2, we have

$$AP = P\Lambda \tag{5.3}$$

for $P \equiv [p_1\ p_2]$ and

$$\Lambda \equiv \begin{pmatrix} \lambda_1 & 0 \\ 0 & \lambda_2 \end{pmatrix}.$$

Then, since P is constant and invertible (due to the linear independence of its columns), the nonsingular change of variables

$$z \equiv Pw \tag{5.4}$$

converts the given system (5.2) to the diagonal system

$$\dot{w} = P^{-1}\dot{z} = P^{-1}Az = \Lambda w \tag{5.5}$$

since $\Lambda = P^{-1}AP$. Moreover, (5.5) has the general (complex) solution

$$w_i(t) = e^{\lambda_i t}c_i, \quad i = 1 \text{ and } 2 \tag{5.6}$$

(for arbitrary constants c_i), which is somewhat more convenient to use than the corresponding general solution

$$z(t) = e^{\lambda_1 t}p_1c_1 + e^{\lambda_2 t}p_2c_2 \tag{5.7}$$

for (5.2). Note that when $\lambda_1 = \overline{\lambda}_2 = \alpha + i\beta$ is complex (so $p_1 = \overline{p}_2$ as well), linearly independent real solutions for (5.2) can be obtained as the real and imaginary parts of the complex solutions. The factorization $A = P\Lambda P^{-1}$ implies that $e^{At} = Pe^{\Lambda t}P^{-1}$ and the solution is $z(t) = e^{At}z(0)$. Since A and Λ are similar matrices, one calls (5.4) a similarity transformation.

Basic linear algebra guarantees that distinct eigenvalues of A will have corresponding linearly independent eigenvectors, so we need to modify the preceding solution procedure when A has a repeated real eigenvalue $\lambda = \frac{1}{2}(a + d)$, but only one corresponding linearly independent eigenvector p_1. Then, we shall rely on linear algebra to guarantee the existence of a (nonunique) generalized eigenvector p_2 defined by $(A - \lambda I)p_2 = p_1$, which is necessarily linearly independent of p_1. Since $Ap_1 = \lambda p_1$ and $Ap_2 = p_1 + \lambda p_2$, the nonsingular generalized modal matrix $P \equiv (p_1\ p_2)$ will satisfy

$$AP = PJ \quad \text{for } J \equiv \begin{pmatrix} \lambda & 1 \\ 0 & \lambda \end{pmatrix} \tag{5.8}$$

being the so-called Jordan canonical (or normal) form for the matrix A. Now $w = P^{-1}z$ will satisfy the triangular system $\dot{w} = Jw$ or, componentwise,

$$\begin{cases} \dot{w}_1 = \lambda w_1 + w_2 \\ \dot{w}_2 = \lambda w_2. \end{cases} \tag{5.9}$$

The general solution of (5.9) is readily obtained as

$$\begin{cases} w_1(t) = e^{\lambda t}(k_1 + k_2 t) \\ w_2(t) = e^{\lambda t}k_2, \end{cases} \tag{5.10}$$

and therefore

$$w(t) = e^{Jt}w(0) = e^{\lambda t}\begin{pmatrix} 1 & t \\ 0 & 1 \end{pmatrix} k \text{ for } w(0) = k = \begin{pmatrix} k_1 \\ k_2 \end{pmatrix}.$$

Thus, the general solution for (5.2) in this case is

$$z(t) = Pw = e^{\lambda t}P\begin{pmatrix} 1 & t \\ 0 & 1 \end{pmatrix} k = e^{\lambda t}(p_1 \; p_1 t + p_2)k, \tag{5.11}$$

with $P = (p_1 \; p_2)$ and $k = P^{-1}z(0)$ arbitrary.

We will find that much useful information about solutions of (5.1) and (5.5) or (5.9) can be gained by studying solutions geometrically as evolving curves in the (w_1, w_2) or the (x, y) *phase plane* (parameterized by the independent variable t). Moreover, the long-time behavior of solutions will be primarily determined by the signs of the eigenvalues λ of A or of their real parts whenever the eigenvalues are complex conjugates. Note that the matrix A has eigenvalues

$$\lambda_{1,2} = \frac{1}{2}(a+d) \pm \sqrt{\frac{1}{4}(a-d)^2 + bc}, \tag{5.12}$$

which are the roots of the characteristic polynomial $|A - \lambda I| = \lambda^2 - (a+d)\lambda + (ad - bc) = 0$. Moreover, when A has two linearly independent eigenvectors p_1 and p_2, (5.7) implies that the general solution of (5.1) has the form

$$\begin{cases} x(t) = e^{\lambda_1 t}p_{11}c_1 + e^{\lambda_2 t}p_{12}c_2 \\ y(t) = e^{\lambda_1 t}p_{21}c_1 + e^{\lambda_2 t}p_{22}c_2, \end{cases} \tag{5.13}$$

where $p_i \equiv \begin{pmatrix} p_{1i} \\ p_{2i} \end{pmatrix}$ for $i = 1$ and 2. In the exceptional case, when we need a generalized eigenvector p_2, (5.11) shows that we will instead have the general solution

$$\begin{cases} x(t) = e^{\lambda t}(p_{11}c_1 + (p_{11}t + p_{12})c_2) \\ y(t) = e^{\lambda t}(p_{21}c_1 + (p_{21}t + p_{22})c_2). \end{cases} \tag{5.14}$$

The simpler solution of the transformed problem for $w(t) = \begin{pmatrix} w_1(t) \\ w_2(t) \end{pmatrix}$ follows by replacing $P = \begin{pmatrix} p_{11} & p_{12} \\ p_{21} & p_{22} \end{pmatrix}$ by $I = \begin{pmatrix} 1 & 0 \\ 0 & 1 \end{pmatrix}$ in these expressions. A unique solution z of (5.2) will be specified by the initial values $x(0)$ and $y(0)$, which determine both constants c_1 and c_2. This will result in a unique curve for the solution's phase-plane trajectory $(x(t), y(t))$ as the parameter t varies.

We shall call the origin, $z = \begin{pmatrix} x \\ y \end{pmatrix} = 0$, a *rest* (or equilibrium) *point* of the system (5.1), since any solution of $\dot{z} = 0$ starting at $z = 0$ stays there for all time t. We will now classify the rest points of (5.2) according to the nature of the eigenpairs of A, first noting that $z = 0$ is the only rest point of $\dot{z} = Az$ whenever the state matrix A is nonsingular, as we shall assume. (When A is singular, the planar dynamics is less than two dimensional, so that situation is of considerably less interest.)

The five cases we need to consider are:

1 We will call the rest point $\begin{pmatrix} x \\ y \end{pmatrix} = 0$ a (stable or) *node* of the linear system (5.1) if both eigenvalues $\lambda_{1,2}$ of A are *real, unequal, and of the same sign*. In the (*asymptotically*) *stable* situation when $\lambda_2 < \lambda_1 < 0$, all solutions (5.13) of (5.1) tend to the trivial rest point as $t \to \infty$. Moreover, all solutions with $c_1 \neq 0$ satisfy

$$\frac{y}{x} \to \frac{p_{21}}{p_{11}} \quad \text{as} \quad t \to \infty.$$

Thus, the limiting behavior is determined by the eigenvectors p_1 corresponding to the least stable eigenvalue λ_1. On the exceptional trajectory, we instead have

$$\frac{y}{x} = \frac{p_{22}}{p_{12}}.$$

A typical *phase-plane portrait* is shown in Figure 7.

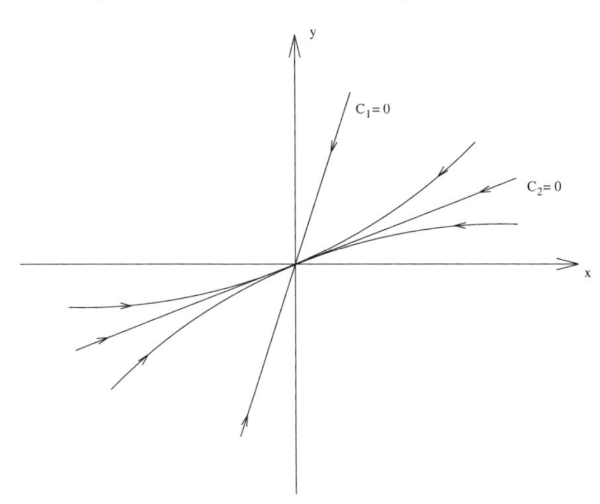

Figure 7. Stable Node

The arrows are used to indicate the direction of motion as t increases. We note that a unique trajectory can be drawn through any point. Indeed, because initial value problems have unique solutions, curves cannot cross (except at the origin, where there is no motion). In the case when $0 < \lambda_1 < \lambda_2$, the arrows on the trajectories must be reversed, and the origin is said to be a *repulsive* rest point as $t \to \infty$ (though *attractive* as $t \to -\infty$). In the corresponding w_1-w_2 phase-plane, the special directions corresponding to $c_j = 0$ coincide with the coordinate axes $w_j = 0$ for $j = 1$ and 2.

2. When the roots $\lambda_{1,2}$ of the characteristic polynomial are *real and of opposite signs*, say $\lambda_2 < 0 < \lambda_1$, the origin is called a *saddle point*. Nonzero solutions (5.13) with $c_1 = 0$ tend to the origin as $t \to \infty$, whereas those with $c_2 = 0$ move away from the origin as t increases. Moreover, solutions for $c_1 c_2 \neq 0$ tend asymptotically toward the direction of the eigenvector p_1 (corresponding to the dominating eigenvalue) as $t \to \infty$ and toward that of p_2 as $t \to -\infty$. Pictorially, we have

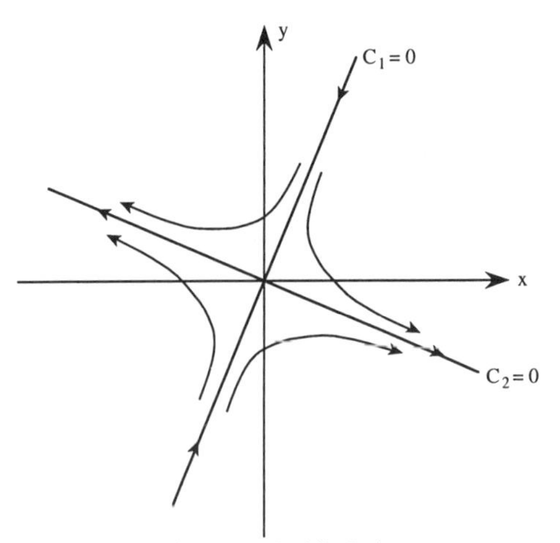

Figure 8. Saddle Point

The origin is not an attractor, either as $t \to \infty$ or $t \to -\infty$, since only special trajectories (with either c_1 or c_2 zero) tend to it in that limit. We shall naturally call such rest points , though recognizing that there is actually a one-dimensional *stable manifold* (with $c_1 = 0$ and c_2 free) and a corresponding one-dimensional (with $c_2 = 0$).

3. If the roots $\lambda_1 = \lambda_2 = \lambda \equiv \frac{1}{2}(a + d) \neq 0$ are *repeated*, but nonzero, we will again call the origin a *node* (stable if $\lambda < 0$, but if $\lambda > 0$). If A has a full set of two linearly independent eigenvectors p_1 and p_2, the solutions

(5.7) satisfy

$$\frac{y}{x} = \frac{p_{21}c_1 + p_{22}c_2}{p_{11}c_1 + p_{12}c_2},$$

i.e., they remain on straight-line paths (which enter the origin as $t \to \infty$ if $\lambda < 0$). The phase portrait is then given by

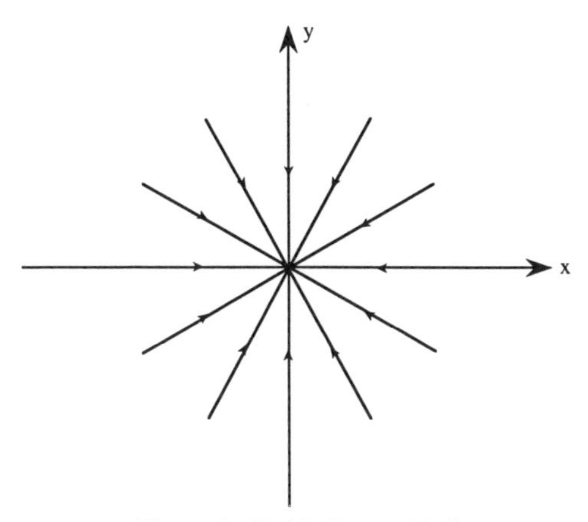

Figure 9. Stable Proper Node

We call such rest points *proper nodes*. Because c_1 and c_2 are arbitrary, there are no distinguished directions in this case.

In the exceptional case when A has only one linearly independent eigenvector $p_1 = \begin{pmatrix} p_{11} \\ p_{21} \end{pmatrix}$, the limiting slope of all solutions (5.14), $\frac{p_{21}}{p_{11}}$, is determined by the eigenvector. Pictorially, then, for $\lambda < 0$ we will have the phase portrait of Figure 10.

The origin is again called a node.

4 When $(a-d)^2 < -4bc$, the roots $\lambda_{1,2}$ of the characteristic polynomial are *conjugate complex pairs* $\alpha \pm i\beta \equiv \frac{1}{2}(a+d) \pm \frac{i}{2}\sqrt{-(a-d)^2 - 4bc}$. We will call the origin a *spiral point* if α is nonzero. It will be *stable* () if $\alpha < (>) 0$. (If $\alpha = 0$, the solution is (bounded and) periodic and the origin is a stable *center* or *sink.*)

If we make the (unmotivated, but effective) nonsingular transformation

$$\begin{pmatrix} u \\ v \end{pmatrix} \equiv Q \begin{pmatrix} x \\ y \end{pmatrix},$$

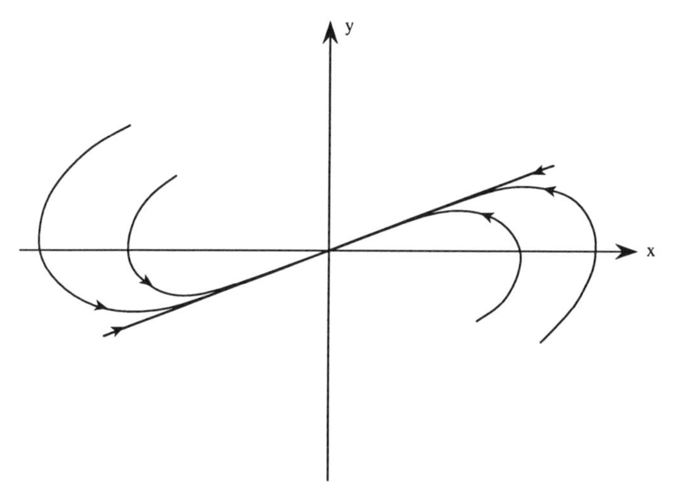

Figure 10. Another Stable Node

for

$$Q \equiv \begin{pmatrix} c & \alpha - a \\ 0 & \beta \end{pmatrix},$$

we obtain

$$\begin{pmatrix} \dot{u} \\ v \end{pmatrix} \equiv QAQ^{-1} \begin{pmatrix} u \\ v \end{pmatrix} = \begin{pmatrix} \alpha & -\beta \\ \beta & \alpha \end{pmatrix} \begin{pmatrix} u \\ v \end{pmatrix}.$$

Then, introducing the *polar coordinates* $u = r\cos\theta$ and $v = r\sin\theta$, $rdr = udu + vdv$ and $r^2 d\theta = udv - vdu$, so we get the simple equivalent, but decoupled, transformed problem

$$\begin{cases} r\dfrac{dr}{dt} = u(\alpha u - \beta v) + v(\beta u + \alpha v) = \alpha r^2 \\ \text{and} \\ r^2\dfrac{d\theta}{dt} = u(\beta u + \alpha v) - v(\alpha u - \beta v) = \beta r^2. \end{cases}$$

Integrating then yields

$$r(t) = e^{\alpha t} r(0) \text{ and } \theta(t) = \beta t + \theta(0)$$

and, thereby, $u(t) + iv(t) = re^{i\theta} = e^{\alpha t} r(0)[\cos(\beta t + \theta(0)) + i\sin(\beta t + \theta(0))]$. For the stable case where $\alpha < 0$, the trajectory tends to the origin of the u-v plane as $t \to \infty$. Moreover, $\theta(t)$ increases monotonically since $\beta > 0$. In the u-v phase plane, the trajectories have a spiral shape. The corresponding orbit is, however, somewhat distorted in the x-y

plane since

$$\begin{pmatrix} x \\ y \end{pmatrix} = Q^{-1} \begin{pmatrix} u \\ v \end{pmatrix} = \begin{pmatrix} \beta & a-\alpha \\ 0 & c \end{pmatrix} \begin{pmatrix} \cos(\beta t + \theta(0)) \\ \sin(\beta t + \theta(0)) \end{pmatrix} e^{\alpha t} r(0).$$

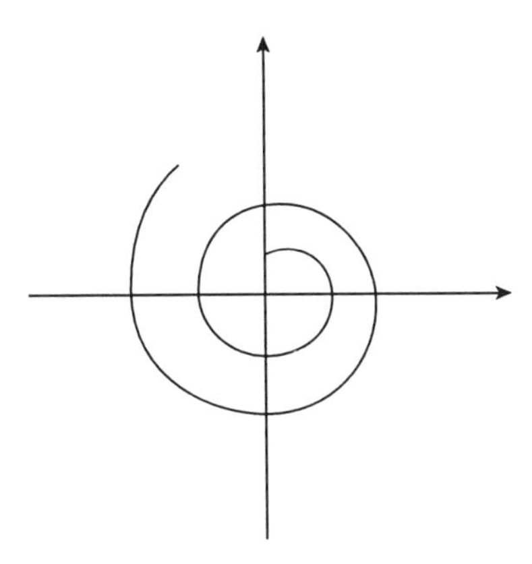

Figure 11. A Stable Spiral

5 When $a = -d$ and $a^2 + bc < 0$, the roots of the characteristic polynomial are the purely imaginary numbers $\lambda_{1,2} = \pm i\beta \equiv \pm\sqrt{a^2 + bc}$. In the u-v plane, we obtain real solutions

$$u(t) = r(0) \cos(\beta t + \theta(0)) \quad \text{and} \quad v(t) = r(0) \sin(\beta t + \theta(0)),$$

of period $2\pi/\beta$, so trajectories lie on circles centered at the origin.

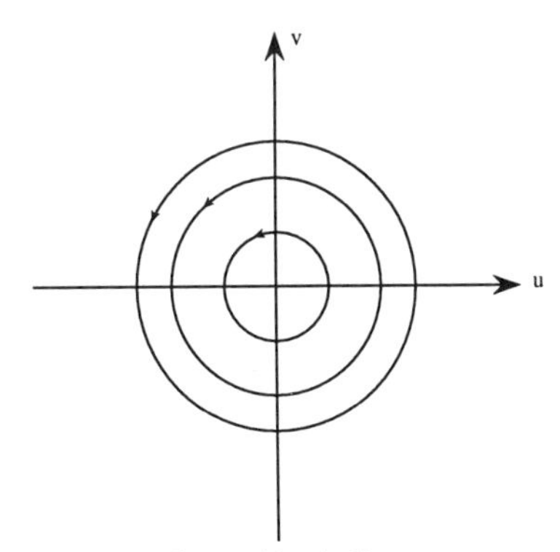

Figure 12. A Center

In the original coordinates, trajectories

$$\begin{pmatrix} x \\ y \end{pmatrix} = \begin{pmatrix} \beta\cos(\beta t + \theta(0)) + (a-\alpha)\sin(\beta t + \theta(0)) \\ c\sin(\beta t + \theta(0)) \end{pmatrix} r(0)$$

will generally be elliptical, but they, too, have period $2\pi/\beta$. Because we can keep $|x(t)|$ and $|y(t)|$ as small as we wish by picking $r(0)$ sufficiently small, we will call the rest point *stable*. However, because nearby trajectories do not tend to the origin as $t \to \infty$, we no longer have the more desirable property of asymptotic stability. One calls such rest points *centers*.

5.2 Using the phase plane for nonlinear problems

If we reconsider the second-order scalar autonomous equation

$$\frac{d^2x}{dt^2} = F\left(x, \frac{dx}{dt}\right),$$

we can study trajectories in the $\left(x, \frac{dx}{dt}\right)$ phase plane by introducing $y = \frac{dx}{dt}$ such that $\frac{dy}{dt} = F(x, y)$. More generally, we can consider nonlinear planar systems

$$\begin{cases} \frac{dx}{dt} = P(x, y) \\ \frac{dy}{dt} = Q(x, y), \end{cases} \tag{5.15}$$

for any smooth scalar functions P and Q. By eliminating t, we are left with the scalar first-order equation

$$\frac{dy}{dx} = \frac{Q(x, y)}{P(x, y)} \tag{5.16}$$

for $y(x)$ or $x(y)$. The "time" t will then parameterize motion along trajectories in the (x, y) phase plane. Indeed, the time elapsed along a trajectory can be obtained by integrating $\frac{dx}{P(x,y)}$ or $\frac{dy}{Q(x,y)}$. Since motion is stationary at rest points (x_0, y_0), where $P(x_0, y_0)$ and $Q(x_0, y_0)$ are simultaneously zero, we will naturally be concerned about the (long-time) evolution of trajectories that begin near such points. If we expand $P(x, y)$ and $Q(x, y)$ in their Taylor series expansions, for example,

$$\begin{aligned} P(x, y) = {} & P_x(x_0, y_0)(x - x_0) + P_y(x_0, y_0)(y - y_0) \\ & + \frac{1}{2}P_{xx}(x_0, y_0)(x - x_0)^2 + P_{xy}(x_0, y_0)(x - x_0)(y - y_0) \\ & + \frac{1}{2}P_{yy}(x_0, y_0)(y - y_0)^2 + \dots, \end{aligned}$$

we obtain the approximate system

$$\begin{cases} \frac{du}{dt} \approx P_x(x_0, y_0)u + P_y(x_0, y_0)v \\ \frac{dv}{dt} \approx Q_x(x_0, y_0)u + Q_y(x_0, y_0)v \end{cases} \tag{5.17}$$

for $u \equiv x - x_0$ and $v \equiv y - y_0$ when we only retain the linear terms. Equation (5.17) can be analyzed near rest points by the methods of the preceding section, provided the coefficient matrix

$$J(x_0, y_0) \equiv \begin{pmatrix} P_x(x_0, y_0) & P_y(x_0, y_0) \\ Q_x(x_0, y_0) & Q_y(x_0, y_0) \end{pmatrix}$$

(i.e., the Jacobian) is nonsingular. We may further hope that the *local* behavior of solutions to the nonlinear system (5.15) and to its linear approximation (5.17) will be *qualitatively* similar, even as $t \to \pm\infty$. A still better approximation would result if one retained quadratic terms in the Taylor series expansions. The detailed study of all such possibilities involves quite recent research which has, to some extent, been a speciality of Chinese mathematicians.

Example 1 Consider the nonlinear pendulum, whose angle of deflection θ satisfies

$$\frac{d^2\theta}{dt^2} + k^2 \sin\theta = 0. \tag{5.18}$$

If we introduce the angular velocity $\omega = \frac{d\theta}{dt}$, the resulting first-order system

$$\begin{cases} \frac{d\theta}{dt} = \omega \\ \frac{d\omega}{dt} = -k^2 \sin\theta \end{cases}$$

will have rest points at $(\theta_0, \omega_0) = (n\pi, 0)$, $n = 0, \pm 1, \pm 2, \dots$. Since θ increases with t for $\omega > 0$, motion will be to the right (left) in the upper (lower) half of the θ-ω plane.

Trajectories in this phase plane satisfy the separable first-order equation

$$\frac{d\omega}{d\theta} = -\frac{k^2 \sin\theta}{\omega}.$$

Integration implies that

$$\frac{\omega^2}{2} - k^2 \cos\theta = \frac{c}{2}.$$

This relation can be interpreted as a conservation of energy statement along each orbit; the total energy $c/2$ is a constant ($\geq -k^2$) that is uniquely specified by any prescribed initial point $[\theta(0), \omega(0)]$. Indeed,

$$\omega(\theta) = \pm\sqrt{2k^2 \cos\theta + c},$$

with the sign selected being that of $\omega(0)$. For $c = 2k^2$, the corresponding trajectory passes through the rest points $[(2n+1)\pi, 0]$ and encircles the remaining ones. (When we take time t into account, passage through a rest point takes an infinite amount of time.) For $c > 2k^2$, one never attains zero velocity, so θ increases or decreases indefinitely, depending on the sign of ω. For $c < 2k^2$, phase-plane trajectories encircle the rest points $(2n\pi, 0)$ indefinitely (and the resulting trajectory passes through its initial point periodically). We have the following portrait (for a typical constant k), which is repeated with 2π-periodicity in θ:

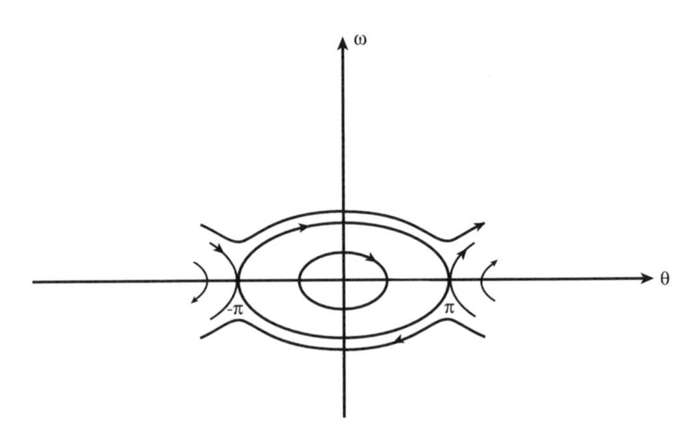

Figure 13. Pendulum's Phase Portrait

Because trajectories beginning near the rest points $(2n\pi, 0)$ remain nearby for all time, we will call these rest points *stable centers*; the remaining rest points, $(m\pi, 0)$ for m odd, will be *() saddles*. This corresponds physically to the clear distinction between a pendulum held straight down or straight up, the latter rest point allowing no tolerance for error!

The nonlinear pendulum represents a special case of the scalar second-order nonlinear *conservation equation*

$$\frac{d^2x}{dt^2} + F(x) = 0 \tag{5.19}$$

for a general function $F(x)$, whose solutions $x(t)$ are naturally parameterized by the initial values $x(0)$ and $\frac{dx}{dt}(0)$. If we introduce the "velocity" $v = \frac{dx}{dt}$ and the "potential energy"

$$V(x) = \int_{x(0)}^{x} F(s)ds, \tag{5.20}$$

we can multiply (5.19) by $\frac{dx}{dt}$ and integrate to find that the "total energy"

$$\frac{1}{2}v^2 + V(x) = h = \frac{1}{2}v^2(0) \tag{5.21}$$

is conserved on every trajectory. Trajectories in the x-v phase plane can be efficiently parameterized using their total energy h, rather than their initial values. Since

$$v(x) = \pm\sqrt{2(h - V(x))}, \tag{5.22}$$

one can determine the velocity $v(x)$ for any x by plotting the potential energy V versus x. For each h exceeding the minimum of V, we determine the range of x values corresponding to (the physically meaningful) real velocities v. Pictorially, we have

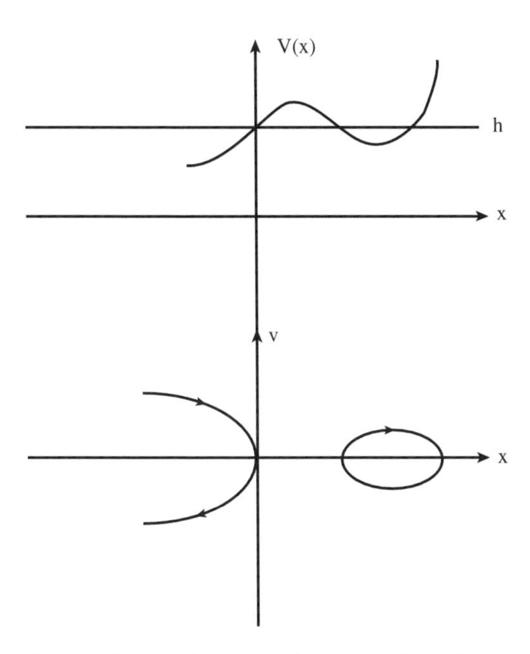

Figure 14. Determining the Phase Portrait from the Potential Energy Profile

Recall that $\frac{dx}{dt} = v$ is positive in the upper half of the x-v plane and negative in the lower half. Rest points $(x_0, 0)$ therefore only occur on the x axis when

$F(x_0) = 0$. Moreover, it is straightforward (on both mathematical and physical grounds) that they are *stable centers* when $F'(x_0) > 0$ (so the rest point occurs at a minimum of the potential energy V) and they are *saddle points* when x_0 is a corresponding maximum of $V(x)$. Much insight can be achieved by considering *separatrices*, i.e. orbits corresponding to energy levels which separate, say, periodic motion (closed planar trajectories) and, e.g., wavy motion.

Example 2 Consider the two-point boundary value problem

$$y'' = e^{-2y}, \quad y(0) = y(T) = 0, \tag{5.23}$$

which is similar to ones occurring in *chemical kinetics* and *combustion theory* (where the exponential term represents so-called Arrhenius kinetics). The differential equation can be integrated once, after being multiplied by y', to yield the conservation law

$$(y')^2 + e^{-2y} = (y'(0))^2 + 1 = a^2.$$

Since $y'(0)$ is not specified, the parameter $a = \sqrt{(y'(0))^2 + 1}$ exceeds $|y'(0)|$ and is also not less than 1. Introducing $v = y'$ provides the phase-plane system

$$\begin{cases} y' = v \\ v' = e^{-2y} \end{cases}$$

and the separable equation $\frac{dv}{dy} = \frac{e^{-2y}}{v}$ to describe motion along trajectories. (Its solution $v^2 + e^{-2y} = a^2$, of course, coincides with the conservation law.)

Since $e^{-2y} = y'' = v'$, we obtain the separable equation

$$v' = a^2 - v^2$$

for v. Moreover, $v' = e^{-2y} > 0$ implies that v will increase monotonically with t. Since y must vary from 0 back to 0 as t varies from 0 to T, $y' = v$ implies that $v(T) = \sqrt{a^2 - 1} = -v(0)$ is positive. The conservation of $v^2 + e^{-2y}$ leaves only $a = \sqrt{v^2(0) + 1}$ (or, equivalently, $v(0)$) to be specified. In the phase plane, we will have the portrait of Figure 15.

Our solution is thereby numerically reduced to a shooting problem to find the constant a (i.e., guess a, determine the resulting endvalue $y(T)$, adjust a, and repeat). First note that the separable equation for v implies that

$$\frac{v(t) + a}{v(t) - a} = \left(\frac{v(0) + a}{v(0) - a} \right) e^{2at}.$$

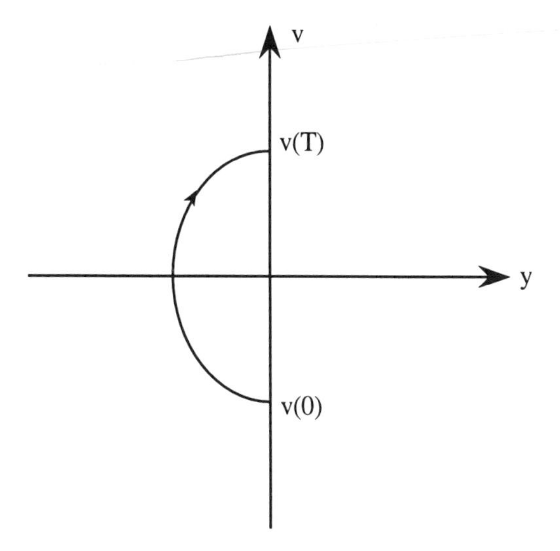

Figure 15. Phase Plane Portrait for Example 2

Because $v(T) = -v(0)$, a satisfies the transcendental equation $e^{aT} = \frac{a-v(0)}{a+v(0)} = (a - v(0))^2$. Taking the square root gives $e^{aT/2} = a + \sqrt{a^2 - 1} = e^{\cosh^{-1} a}$. This provides the fairly simple equation

$$\cosh\frac{aT}{2} = a \quad \text{or} \quad \frac{aT}{2} = \cosh^{-1} a$$

for a, parameterized by the interval length T. A plot of this function is shown in Figure 16.

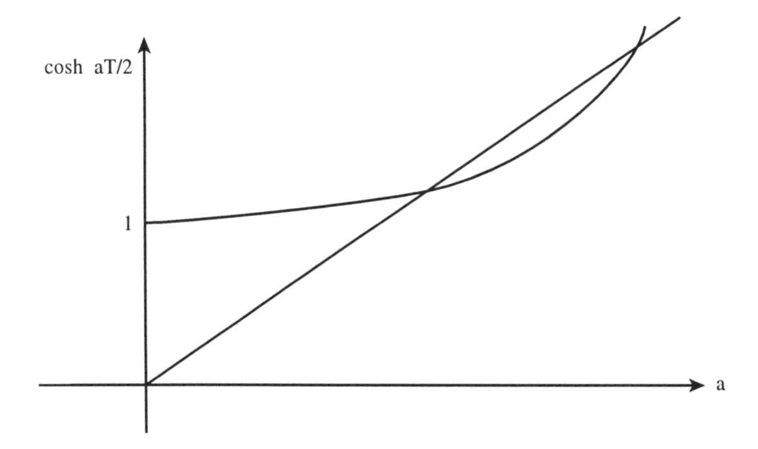

Figure 16. Determining roots a and corresponding solutions of Example 2.

If we consider various lengths T, we find that there is a threshold T_0 such that we get the solution $a = 1$ for $T = 0$, two solutions a for $0 < T < T_0$, one solution for $T = T_0$, and no solutions for $T > T_0$. The complete loss of solutions at T_0, where the curves intersect tangentially, would be especially difficult to detect numerically.

The interval length T is, therefore, very critical. For $0 < T < T_0$, the two interesting solutions (corresponding to each of the two roots a) determine

$$v(t) = \frac{a(e^{a(2t-T)} - 1)}{e^{a(2t-T)} + 1} = a \tanh\left(a\left(t - \frac{T}{2}\right)\right).$$

This provides $e^{-2y} = a^2 - v^2 = a^2 \text{sech}^2\,(a(t - \frac{1}{2}T))$, i.e., the two interesting solutions

$$y(t) = \ln\left[\frac{1}{a}\cosh\left(a\left(t - \frac{T}{2}\right)\right)\right].$$

Solving nonlinear two-point boundary value problems is clearly a (fascinating) challenge, making the more familiar solution of initial value problems look like child's play!

5.3 Stability for higher-dimensional systems

Recall (equation (4.22)) that the constant linear homogeneous system $\dot{x} = Ax$ has the unique solution

$$x(t) = e^{At}x(0) = \left(\sum_{i=1}^{r} e^{\lambda_i t} P_i \sum_{j=0}^{m_i - 1} (A - \lambda_i I)^j \frac{t^j}{j!}\right) x(0)$$

when the characteristic polynomial $p(\lambda) = \det(A - \lambda I) = 0$ has the factorization $p(\lambda) = (\lambda_1 - \lambda)^{m_1}(\lambda_2 - \lambda)^{m_2} \ldots (\lambda_r - \lambda)^{m_r}$ and each P_i is a matrix that projects any vector onto the m_i-dimensional generalized eigenspace of A corresponding to the eigenvalue λ_i (cf. (4.20)). This representation of solutions has three immediate consequences:

1 If all the roots λ_i of $p(\lambda)$ have negative real parts, then all solutions $x(t)$ are linear combinations of terms $t^j e^{\lambda_j t}$, so they tend (exponentially) to zero as $t \to \infty$. We then call the origin an *asymptotically stable* rest point.

2 If all the roots λ_i of $p(\lambda)$ satisfy $\Re\lambda_i \leq 0$ and any purely imaginary roots are *simple*, then all solutions $x(t)$ are bounded as $t \to \infty$, since they are linear combinations of decaying terms $t^j e^{\lambda_j t}$ with $\Re\lambda_j < 0$ or bounded terms $e^{\pm i\beta t}$ with β real (possibly zero). We then call the rest point $x = 0$ stable, but not asymptotically stable.

3 If any root λ_i of $p(\lambda)$ has a positive real part or if a purely imaginary root is multiple, then (depending on $x(0)$) some solutions $x(t)$ might be unbounded as $t \to \infty$ (like $t^j e^{\lambda_i t}$ for $\Re\lambda_i > 0$ or like $t^j e^{i\beta t}$ for $j > 0$ and β real). We call this *instability*.

Example 3 The trivial solution of

$$\dot{x} = \begin{pmatrix} -1 & 0 & 0 \\ 0 & -2 & 0 \\ 0 & 0 & -3 \end{pmatrix} x$$

is naturally called asymptotically stable. Indeed, we shall call a square matrix A stable if all its eigenvalues lie *strictly* in the left half plane, causing all solutions of the corresponding differential system $\dot{x} = Ax$ to tend to zero as $t \to \infty$.

Example 4 The trivial solution of

$$\dot{x} = \begin{pmatrix} 0 & 0 & 0 \\ 0 & -1 & 0 \\ 0 & 0 & -2 \end{pmatrix} x$$

is stable, but not asymptotically stable, because the state matrix has a single zero eigenvalue and the constant solutions

$$x(t) = \begin{pmatrix} 1 \\ 0 \\ 0 \end{pmatrix} c_1$$

do not decay as t increases.

Example 5 The trivial solution of

$$\dot{x} = \begin{pmatrix} 0 & \delta \\ 0 & 0 \end{pmatrix} x$$

is stable if $\delta = 0$ and is otherwise. In both cases, the trivial eigenvalue $\lambda = 0$ has multiplicity two. In the first case, however, there are two linearly independent nullvectors (i.e., eigenvectors corresponding to $\lambda = 0$) and all solutions remain constant as $t \to \infty$. In the second case, unbounded solutions, growing like multiples of t, occur, unless we restrict $x(0)$ to being a multiple of the nullvector $\begin{pmatrix} 1 \\ 0 \end{pmatrix}$. Then, however, computed solutions will ultimately become unreliable (due to the buildup of roundoff error in the orthogonal space spanned by the generalized nullvector $\begin{pmatrix} 0 \\ 1 \end{pmatrix}$). (Readers are encouraged to experiment with their favorite software and watch the discrepancy build up, moving the computed solution away from the true constant solution.)

Example 6 The rest point $x(t) \equiv 0$ for the *conditionally stable* linear system

$$\dot{x} = \begin{pmatrix} 1 & 0 & 0 \\ 0 & -1 & 0 \\ 0 & 0 & -2 \end{pmatrix} x$$

is because it has the positive eigenvalue 1, though solutions in the two-dimensional *stable eigenspace* spanned by

$$\begin{pmatrix} 0 \\ 1 \\ 0 \end{pmatrix} \text{ and } \begin{pmatrix} 0 \\ 0 \\ 1 \end{pmatrix}$$

(eigenvectors corresponding to the system's stable eigenvalues -1 and -2) decay exponentially as $t \to \infty$. To keep solutions bounded, we must avoid initial values $x(0)$ with *any* component in the eigenspace determined by the eigenpair

$$\left(1, \begin{pmatrix} 1 \\ 0 \\ 0 \end{pmatrix}\right).$$

The smallest contamination from this orthogonal subspace (as is inherent in numerical computation) will ultimately imply unboundedness.

Let us now consider a nonlinear vector initial value problem

$$\dot{x} = f(x, t), x(t_0) = x_0, \tag{5.24}$$

where f and f_x are continuous for all x and $t \geq t_0$, and let us denote the unique solution by

$$x(t) = x(t; x_0, t_0).$$

In many physical situations, it is critical to know how much the solution changes under small perturbations in the vector function f, the initial vector x_0, or the initial point t_0. We, however, will limit attention to solutions $x(t; x_0, t_0)$ and $y(t; y_0, t_0)$ for a y_0 near x_0, noting that the difference $z = y - x$ then satisfies

$$\dot{z} = F(z, t) \equiv f(x(t) + z, t) - f(x(t), t). \tag{5.25}$$

Since $F(0, t) = 0$ for all $t \geq t_0$, (5.25) has the trivial solution $z(t) \equiv 0$ if $z(t_0) = 0$.

We will introduce the following terminology, due to Liapunov (about 1892):

Definition 5.3.1 The solution $z(t) \equiv 0$ of $\dot{z} = F(z, t)$ will be

(i) *stable* if, for all $\epsilon > 0$ and any $t_1 \geq t_0$, there is a $\delta(\epsilon, t_1) > 0$ such that $|z(t_1)| < \delta$ implies that $|z(t)| < \epsilon$ for all $t \geq t_1$.
Pictorially, stability can be defined using the figure:

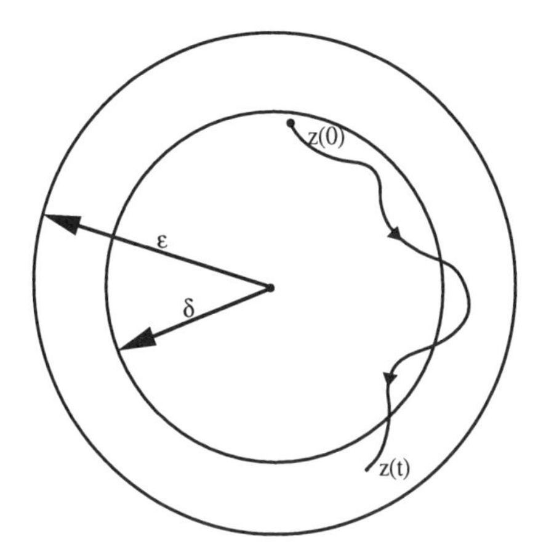

Figure 17. Stability neighborhood

If these conditions are not satisfied, we will call $z(t) \equiv 0$.
Continuing, we shall call $z(t) \equiv 0$

(ii) *uniformly stable* if it is stable and if we can take $\delta(\epsilon, t_1)$ to be independent of t_1.
(iii) *asymptotically stable* if it is stable and if, for some $\tilde{\delta} > 0$, $|z(t_1)| < \tilde{\delta}$ implies that $z(t) \to 0$ as $t \to \infty$.

Numerous stability concepts, with subtle differences, have been used throughout the applied mathematics and engineering literatures (cf. Hahn (1963) and Coppel (1965)). An elementary introduction is given by Sanchez (1968). In the numerical analysis literature, related ideas are referred to as stability, *sensitivity*, and *conditioning*.

Let us now restrict attention to certain linear systems of the form

$$\dot{x} = A(t)x + g(t)$$

when the matrix $A(t)$ and the vector $g(t)$ are both continuous for $t \geq 0$. Then, stability is determined by the homogeneous system (i.e., the state matrix $A(t)$) and by properties of the corresponding fundamental matrix $\Phi(t)$. Indeed, the behavior of solutions as $t \to \infty$ *may* be determined by the limiting matrix $A_\infty = \lim_{t\to\infty} A(t)$. Consider the special homogeneous case

$$\dot{z} = (A_\infty + B(t))z, \tag{5.26}$$

where A_∞ is a stable matrix and where $B(t)$ is continuous for $t \geq 0$ and $B(t) \to 0$ as $t \to \infty$. If we treat the term $B(t)z$ as a nonhomogeneity, we can use variation of parameters to convert the linear system (5.26) to the equivalent linear vector *integral equation*

$$z(t) = e^{A_\infty(t-t_1)}z(t_1) + \int_{t_1}^{t} e^{A_\infty(t-s)}B(s)z(s)ds$$

for $t \geq t_1$. Since all the eigenvalues of A_∞ lie strictly in the left half plane, using any matrix norm, we will have

$$|e^{A_\infty(t-t_1)}| \leq Me^{-\alpha(t-t_1)} \tag{5.27}$$

on $t \geq t_1$ for some positive constants M and α. Using compatible vector norms, we directly obtain the scalar inequality

$$\begin{aligned} |z(t)| &\leq |e^{A_\infty(t-t_1)}|\ |z(t_1)| + \int_{t_1}^{t} |e^{A_\infty(t-s)}|\ |B(s)|\ |z(s)|ds \\ &\leq M\left[e^{-\alpha(t-t_1)}|z(t_1)| + \int_{t_1}^{t} e^{-\alpha(t-s)}|B(s)|\ |z(s)|ds\right] \end{aligned}$$

for $t \geq t_1$. Now, the scalar

$$r(t) \equiv |z(t)|e^{\alpha(t-t_1)}$$

satisfies the *integral inequality*

$$r(t) \leq M\left[r(t_1) + \int_{t_1}^{t} |B(s)|r(s)ds\right] \equiv R(t)$$

for $t \geq t_1$. We will be able to get an upper bound for $r(t)$ on $t \geq t_1$ using a now-classical argument, due to Gronwall. Note that

$$R'(t) = M|B(t)|r(t) \leq M|B(t)|R(t).$$

Since $R(t) \geq 0$, we can divide by R to get the bound $\frac{d}{dt}(\ln R(t)) \leq M|B(t)|$. Upon integrating, we get

$$r(t) \leq R(t) \leq Mr(t_1)e^{M\int_{t_1}^{t}|B(s)|ds} \quad \text{for } t \geq t_1.$$

By the definition of r, this, moreover, guarantees that the vector solution $z(t)$ of (5.26) satisfies

$$|z(t)| \leq M|z(t_1)|e^{-\alpha(t-t_1)+M\int_{t_1}^{t}|B(s)|ds}.$$

Since $B(t) \to 0$ as $t \to \infty$ and B is bounded on any finite interval, there is a positive bound $|B(t)| \leq \gamma$ for all $t \geq t_0$. Further, given any smaller constant $\beta > 0$, there will be a t_2 that depends on β such that $|B(t)| < \beta$ for $t \geq t_2$. Then, for all $t \geq t_1$, $\int_{t_1}^{t}|B(s)|ds \leq \beta(t-t_1)+\gamma(t_2-t_1)$ implies that

$$|z(t)| \leq M|z(t_1)|e^{M\gamma(t_2-t_1)}e^{-(\alpha-M\beta)(t-t_1)}. \tag{5.28}$$

Picking $\beta = \frac{\alpha}{2M} > 0$ finally shows that $|z(t)|$ decays like the exponential $e^{-\frac{\alpha}{2}(t-t_1)}$ for $t \geq t_1$. This demonstrates that our simple hypotheses on A_∞ and $B(t)$ guarantee the asymptotic stability of the trivial solution of (5.26).

We observe that Gronwall's inequality is of considerable independent interest. It is easily understood in the scalar case when the inequalities

$$\dot{r} \leq b(t)r + c(t), \quad r(t_1) \leq x(t_1)$$

and the equality

$$\dot{x} = b(t)x + c(t)$$

directly imply that

$$r(t) \leq x(t) \quad \text{for} \quad t \geq t_1.$$

Example 7 The scalar equation

$$\dot{z} = \left(\frac{2t}{1+t^2}\right)z$$

has the unique solution $z(t) = (1+t^2)z(0)$. Thus, all nontrivial solutions become unbounded as $t \to \infty$. This does not contradict our preceding result for (5.26), because here the limit $A_\infty = \lim_{t\to\infty}\left(\frac{2t}{1+t^2}\right) = 0$ is not strictly stable.

For the nonlinear vector system

$$\dot{z} = Az + H(z,t), \tag{5.29}$$

suppose that the constant state matrix A is stable and, for any vector norm, that for any $\beta > 0$, there is a $\delta(\beta)$ such that

$$|H(z,t)| < \beta|z| \text{ whenever } |z| \leq \delta(\beta). \tag{5.30}$$

This inequality will certainly hold when $\dot{z} - Az$ is at least quadratically small with respect to z near $z = 0$. Variation of parameters implies that $z(t)$ must satisfy the nonlinear integral equation

$$z(t) = e^{A(t-t_1)}z(t_1) + \int_{t_1}^{t} e^{A(t-s)}H(z(s), s)ds.$$

The stability of A implies that $|e^{A(t-s)}| \leq Me^{-\alpha(t-s)}$ for some positive constants M and α. Therefore, the norm $|z|$ satisfies the scalar linear integral inequality

$$|z(t)| \leq M|z(t_1)|e^{-\alpha(t-t_1)} + M\beta \int_{t_1}^{t} |z(s)|e^{-\alpha(t-s)}ds$$

for $t \geq t_1$, presuming $|z(s)| \leq \delta$ for $t_1 \leq s \leq t$. As before, the Gronwall argument implies the upper bound

$$|z(t)| \leq M|z(t_1)|e^{-(\alpha - M\beta)(t-t_1)}, \tag{5.31}$$

so by choosing $\beta = \frac{\alpha}{2M}$ and $|z(t_1)| < \delta/M$, we again show the asymptotic stability of the trivial solution. Our analysis therefore shows that asymptotic stability of the linearized problem $\dot{x} = Ax$ guarantees that of the nonlinear problem (5.29), provided the perturbation H goes to zero faster than linearly near the origin, i.e., that the stable matrix A contains all linear terms in the Maclaurin expansion of $\dot{x}$. This basic result is due to Perron.

Examples 8 and 9 The function $z(t) = \frac{z(t_1)}{\sqrt{1+z^2(t_1)(t-t_1)}}$ satisfies $\dot{z} = -\frac{1}{2}z^3$. The origin is asymptotically stable, since all solutions tend to zero as $t \to \infty$. For the problem $\dot{z} = z^2$, however, the trivial solution is since the solution $z(t) = \frac{z(t_1)}{1-z(t_1)(t-t_1)}$ blows up at $t = t_1 + \frac{1}{z(t_1)}$ whenever $z(t_1) > 0$. The linear approximation $\dot{z} = 0$ provides inadequate information about the stability of these nonlinear systems, because its trivial solution is only stable, rather than not asymptotically stable.

5.4 Liapunov functions

A *Liapunov function* $V(x)$ for the nonlinear vector autonomous system

$$\dot{x} = f(x),$$

which is assumed to have a rest point $\overline{x}$, is a locally continuously differentiable scalar function such that

(i) V is *positive definite*, i.e., $V(\overline{x}) = 0$ and $V(x) > 0$ elsewhere and

(ii) $\dot{V}(x(t)) = V_x(x)f(x) \leq 0$ for $x \neq \overline{x}$.
Here we have used the chain rule, with $\dot{V} \equiv \frac{dV}{dt} = \frac{dV}{dx}\frac{dx}{dt} = V_x f$, for the gradient (row) vector $\frac{dV}{dx}$.

It is relatively easy to prove the following theorem (due to Liapunov, about 1882):

(i) $\overline{x}$ is *stable* if a Liapunov function V exists.
(ii) $\overline{x}$ is *asymptotically stable* if a *strict* Liapunov function exists (satisfying $\dot{V} < 0$ for $x \neq \overline{x}$).
(iii) $\overline{x}$ is if V is positive definite and $\dot{V} > 0$ for $x \neq \overline{x}$.

Although it is sometimes possible to guarantee the existence of Liapunov functions, their explicit construction generally requires unusual skill and experience. Physical interpretations, often in terms of a conserved energy, provide reliable guidelines for determining Liapunov functions as well as understanding (in hindsight) when one has been found. In the case of (asymptotic) stability, the curves $V(x) = C$ for small constants C define a family of *level curves* enclosing the rest point $\overline{x}$. We note that we can sometimes use the minimum of a Liapunov function to estimate the *basin of attraction* (i.e., the set of trajectories that tend to a stable rest point (or *sink*) as $t \to \infty$). Extensions to nonautonomous systems are available, without much complication.

Example 10 Consider the two-dimensional system

$$\begin{cases} \dot{x} = y + \beta x(x^2 + y^2) \\ \dot{y} = -x + \beta y(x^2 + y^2) \end{cases}$$

for which the corresponding linear system, obtained when $\beta = 0$, has the origin as a stable, but not asymptotically stable, rest point. If we introduce the positive-definite function

$$V(x, y) = x^2 + y^2,$$

the derivative $\frac{dV}{dt}$ along solutions $(x(t), y(t))$ satisfies

$$\dot{V} = 2x(y + \beta x(x^2 + y^2)) + 2y(-x + \beta y(x^2 + y^2)) = 2\beta(x^2 + y^2)^2.$$

Thus, $\dot{V}$ is negative definite and Liapunov's theorem guarantees that the origin is asymptotically stable if $\beta < 0$, stable if $\beta = 0$, and if $\beta > 0$. The circular level curves of V are certainly geometrically convenient for understanding the flow.

Example 11 Consider the nonlinear system

$$\begin{cases} \dot{x} = -x + y \\ \dot{y} = -y - x^3 \end{cases}$$

and the positive definite function $V(x, y) = \frac{1}{4}x^4 + \frac{1}{2}y^2$. Along solutions, we will have

$$\dot{V} = x^3(-x + y) + y(-y - x^3) = -x^4 - y^2,$$

so $\dot{V} \leq -2V$ implies that $V(x(t), y(t)) \leq e^{-2t} V(x(0), y(0))$ for $t \geq 0$. This, in turn, implies the exponential decay of all solutions to zero, i.e., asymptotic stability. We, of course, already know this from Perron's theorem because the state matrix

$$A = \begin{pmatrix} -1 & 1 \\ 0 & -1 \end{pmatrix}$$

for the linear approximation is stable.

5.5 Exercises

1 Consider the scalar differential equation

$$\frac{dx}{dt} = (x - 1)(x - 2)(x - 4)^2.$$

(a) What are the rest points?

(b) Determine the range of initial values providing solutions that are attracted to each of them.

2 Suppose all roots of the characteristic polynomial for the scalar constant-coefficient linear differential equation

$$y^{(n)} + a_1 y^{(n-1)} + \ldots + a_n y = 0$$

have negative real parts. Show that every solution of the differential equation tends to zero as $x \to \infty$.

3 (a) Show that all solutions to

$$y'' + a_1 y' + a_0 y = 0$$

decay to zero as $x \to \infty$ if and only if $a_1 > 0$ and $a_0 > 0$.

(b) Show that all solutions to

$$y''' + a_2 y'' + a_1 y' + a_0 y = 0$$

decay to zero as $x \to \infty$ if and only if $a_2 > 0$, $a_1 > 0$, $a_0 > 0$, and $a_1 a_2 > a_0$.

4 Consider the conditionally stable linear system

$$\dot{x} = \begin{pmatrix} B & 0 \\ 0 & C \end{pmatrix} x + \begin{pmatrix} g_1(t) \\ g_2(t) \end{pmatrix},$$

where B is a constant $k \times k$ matrix and C is a constant $(n-k) \times (n-k)$ matrix, such that the eigenvalues of B and of $-C$ all have negative real parts and that the k-vector g_1 and the $(n-k)$-vector g_2 are bounded and continuous on $-\infty < t < \infty$.

(a) Show that the entries of e^{Bt} all decay like $e^{-\lambda t}$, for some $\lambda > 0$, as $t \to \infty$.

(b) Show that

$$x(t) = \begin{pmatrix} \int_{\infty}^{t} e^{B(t-s)} g_1(s) ds \\ -\int_{t}^{\infty} e^{C(t-s)} g_2(s) ds \end{pmatrix},$$

formally satisfies the given system and that it is defined on the whole line $-\infty < t < \infty$.

5 (a) Consider the decoupled two-dimensional linear system

$$\begin{cases} \dot{x}_1 = -x_1 \\ \dot{x}_2 = -2x_2 \end{cases}$$

for $t \geq 0$. For what initial values $x_1(0)$ and $x_2(0)$ will

$$2 \int_0^{\infty} (x_1^2(s) + 2x_2^2(s)) ds = 1?$$

(b) Consider the system

$$\dot{x} = -Px$$

for an n-vector x and a stable, symmetric matrix $-P$. For what initial vectors $x(0)$ will

$$2 \int_0^{\infty} x^T(s) P x(s) ds = 1?$$

(Note: Here, T denotes the transpose. The transpose of a product of matrices is the product of the transposed matrices in the reversed order. $P = P^T$, due to symmetry.)

6 Suppose $y(x)$ and $z(x)$ both satisfy the scalar equation

$$y' = f(y, x),$$

where f satisfies $|f(y, x) - f(z, x)| \leq L|y - z|$ for some positive constant L. Introducing

$$\sigma(x) = (y - x)^2,$$

obtain an upper bound for $\sigma'(x)$, and thereby one for $\sigma(x)$, on $x \geq 0$, that depends only on x, $\sigma(0)$, and L.

7 Consider the *predator-prey* system

$$\begin{cases} x' = (2 - y)x \\ y' = (x - 2)y. \end{cases}$$

(a) Determine the two rest points of the system.

(b) Suppose that $x(0) > 2$ and $y(0) > 2$. Show that $x(t) \leq e^{-(y(0)-2)t}x(0)$ until $x(t) = 2$.

(c) Show that

$$H(x, y) = x - 2\ln x + y - 2\ln y$$

is constant on trajectories.

(d) Explain why the nontrivial rest point is stable.

8 Consider the nonlinear two-dimensional system

$$\begin{cases} p' = q^3 \\ q' = p \end{cases}$$

in the first quadrant of the phase plane. By introducing $V = pq$, show that the origin is an rest point.

9 Consider the nonlinear vector system

$$\dot{z} = Az + H(z),$$

where H is (at least) quadratic in z. By introducing

$$V(z) = z'Bz,$$

where $B = B'$ is a positive definite constant matrix and the prime denotes transposition, show that the origin is asymptotically stable provided

$$A'B + BA \quad \text{is negative definite.}$$

10 For the *Liénard equation*

$$\frac{d^2x}{dt^2} + f(x)\frac{dx}{dt} + g(x) = 0,$$

introduce $y = \frac{dx}{dt} - F(x)$, where $F(x) \equiv \int_0^x f(s)ds$, and show that (i) $V(x, y) = \frac{1}{2}y^2 + G(x)$, for $G(x) = \int_0^x g(s)ds$, is a Liapunov function when $g(x)F(x) \geq 0$ and $G(x) \geq 0$ near $x = 0$ and (ii) the origin is an asymptotically stable rest point of the x-y (*Liénard*) plane provided $g(x)F(x) > 0$ for $x \neq 0$.

11 Consider the planar system

$$\dot{z} = Az \text{ for } A = \begin{pmatrix} -\alpha & 1 \\ -1 & -1 \end{pmatrix}.$$

Determine the types of rest points obtained as the parameter α varies.

12 One way to study the scalar nonlinear equations

$$\frac{dx}{dy} = \frac{Ax + By}{Cx + Dy}$$

is to introduce t as a parameter and to solve the constant linear system

$$\frac{dz}{dt} = \mathcal{A}z$$

for $z = \begin{pmatrix} x \\ y \end{pmatrix}$ and $\mathcal{A} = \begin{pmatrix} A & B \\ C & D \end{pmatrix}$. Does such a parameterization represent a step forward?

13 Show that solutions of the nonlinear system

$$\begin{cases} \frac{dx}{dt} = e^x - 1 \\ \frac{dy}{dt} = ye^x \end{cases}$$

conserve the ratio $y/(e^x - 1)$ as t evolves. How can t be eliminated in terms of x?

14 (a) Show that the equation

$$\frac{d^2u}{dt^2} + u - 2u^3 = 0$$

for the displacement of a nonlinear spring has the conserved "first integral"

$$v^2 + u^2 - u^4 = A^2 - A^4,$$

where $v = \frac{du}{dt}$ and $\pm A$ is the maximum displacement.

(b) Show that the displacement is given by

$$u = A\mathrm{sn}(\sqrt{1 - A^2}\,t + c)$$

for some constant c, where the *elliptic sine function* sn is defined by

$$\mathrm{sn}^{-1}(x) = \int_0^x \frac{dz}{\sqrt{(1 - z^2)(1 - k^2 z^2)}}$$

and k is a parameter within (0, 1).

15 The linear initial value problem

$$\begin{cases} \dot{x} = -y, & x(0) > 0 \\ \dot{y} = -4x, & y(0) > 0 \end{cases}$$

is an example of Lanchester's model for conflict. If $y(0) > 2x(0)$, show that adversary y wins, i.e., that the component x becomes zero first.

16 (a) If polar coordinates $(x, y) \equiv (r \cos\theta, r \sin\theta)$ are introduced, show that the equation

$$\frac{dy}{dx} = f(x, y)$$

is converted to

$$\frac{dr}{d\theta} = r \left[\frac{\cos\theta - \sin\theta f(r\cos\theta, r\sin\theta)}{\sin\theta - \cos\theta f(r\cos\theta, r\sin\theta)} \right].$$

(b) For a stable center, we had

$$\frac{dy}{dx} = \frac{cx + dy}{ax + by}$$

with $\alpha = \frac{1}{2}(a + d) > 0$ and $\beta = \sqrt{-bc - \frac{1}{4}(a - d)^2}$ real and positive. Use (a) to show that

$$\frac{dr}{d\theta} = \frac{\alpha r}{\beta}.$$

17 Use uniqueness to determine the solution to the initial value problem

$$\begin{cases} \frac{dx}{dt} = 3e^x + y^3 - 4y, & x(0) = 0 \\ \frac{dy}{dt} = 2y^2 - xy^3 - 2e^x, & y(0) = 1. \end{cases}$$

18 Solve the nonlinear phase-plane system

$$\begin{cases} \frac{dx}{dt} = y \\ \frac{dy}{dt} = y^2. \end{cases}$$

19 Show that the unit circle $x^2+y^2=1$ is a trajectory of the nonlinear system

$$\frac{dx}{dt} = -y + (x-y)(x^2+y^2-1)$$
$$\frac{dy}{dt} = x + (x+y)(x^2+y^2-1).$$

Then show that trajectories outside the unit circle move away from the origin as t increases, whereas those inside the circle approach it.

20 Find a conserved quantity along solutions of the phase-plane system

$$\frac{dx}{dt} = x\cos y$$
$$\frac{dy}{dt} = -\sin y + x^2.$$

21 Consider the nonlinear spring-mass system, where the deflection x satisfies the initial value problem

$$\ddot{x} + 2x^3 = 0, \quad x(0) = x_0 > 0, \quad \dot{x}(0) = 0.$$

(a) Find $t(x)$.

(b) Use phase-plane arguments to show that the solution is periodic and that its period T satisfies $x(T/4) = 0$.

(c) Show that $T = \dfrac{4}{x_0}\displaystyle\int_0^1 \frac{ds}{\sqrt{1-s^4}}$.

22 The space-charge field φ for a parallel plate capacitor satisfies

$$\frac{d^2\varphi}{dx^2} + \frac{c}{\epsilon\sqrt{\varphi}} = 0, \quad \varphi(0) = \varphi'(0) = 0 \text{ and } \varphi(1) = V$$

for a given voltage V and a dielectric constant ϵ (not necessarily small). Determine the potential c and show that $\varphi(x) = x^{4/3}V$.

23 Use polar coordinates to show that the origin is a spiral point for the nonlinear system

$$\begin{cases} \dot{x} = -y - x\sqrt{x^2+y^2} \\ \dot{y} = x - y\sqrt{x^2+y^2}, \end{cases}$$

but a center for the corresponding linear system.

24 Suppose the system

$$\begin{cases} \dot{N}_1 = [a_1 - d_1(bN_1 + cN_2)]N_1 \\ \dot{N}_2 = [a_2 - d_2(bN_1 + cN_2)]N_2 \end{cases}$$

models the populations N_1 and N_2 of two species competing for a common food supply. Classify the system's equilibrium points. Show that if $a_1 d_2 > a_2 d_1$, the second species dies out and the population of the first species approaches a limiting size. (This is known as Volterra's exclusion principle)

25 The equation for the central orbit of a planet is

$$\frac{d^2u}{d\theta^2} + u = \alpha + \epsilon u^2,$$

where $u = \frac{1}{r}$ and r and θ are polar coordinates for the location of a planet in the plane of its motion. The term ϵu^2 is the *Einstein correction* and α and ϵ are positive constants, ϵ being very small. Find the equilibrium point corresponding to this small perturbation of the Newtonian orbit. Show that this equilibrium is a center in the $(u, \frac{du}{d\theta})$ plane, according to the linear approximation. Confirm that this is also true for the nonlinear equation using methods for conservative systems.

26 Consider the conservative system described by

$$\ddot{x} + x - \frac{1}{3}x^3 = 0.$$

Find the rest points, plot the potential energy, and sketch the phase-plane portrait. Determine the eigenvalues of the linearized problem at the first critical point to the right of the origin. Compare results obtained with those for $\ddot{x} + \sin x = 0$.

27 Consider the system

$$\begin{cases} \dot{x} = -y^3 \\ \dot{y} = x^3 \end{cases}$$

and evaluate the function $V(t) = x^4 + y^4$ along solutions. Using the basic stability definitions, show why the origin is a stable, but not an asymptotically stable, rest point of the system. (It might, therefore, be called neutrally stable.)

28 Consider the system

$$u' = (-A + \gamma(x^2 + y^2 + z^2)I)u$$

for $u = \begin{pmatrix} x \\ y \\ z \end{pmatrix}$, $A = \begin{pmatrix} \alpha & \beta & 0 \\ -\beta & \alpha & 0 \\ 0 & 0 & \alpha \end{pmatrix}$, and positive constants α, β, and γ.

Introduce a natural candidate for V as a potential Liapunov function and determine the stability of the rest point at the origin.

29 Consider the nonlinear spring described by

$$x'' + cx' + k \sin x = 0$$

for positive constants c and k. Use the function

$$V = k(1 - \cos x) + \frac{1}{2}(x')^2$$

to determine the stability of the rest point. Do you expect your conclusion to be optimal?

30 If A is a real stable matrix, show that one can take

$$B = \int_0^\infty e^{A's} e^{As} \, ds$$

in exercise 9.

Solutions to Exercises 1–10

1 (a) The derivative $\frac{dx}{dt} = (x-1)(x-2)(x-4)^2$ is zero at the rest points $x = 1, 2,$ and 4.

(b) The derivative $\frac{dx}{dt}$ is positive for $x < 1$ and for $2 < x < 4$, whereas $\frac{dx}{dt}$ is negative for $1 < x < 2$. Thus, any initial value $x(0) < 2$ will correspond to a solution $x(t)$ that is attracted to $x = 1$ as $t \to \infty$, $x(0) = 2$ will imply a constant solution $x(t)$, any $x(0)$ in $2 < x < 4$ will provide a solution $x(t)$ attracted to $x = 4$ as t increases, and any $x(0) > 4$ will give a solution that becomes unbounded as $t \to \infty$.

2 Suppose all roots λ of the characteristic polynomial $\lambda^n + a_1\lambda^{n-1} + \ldots + a_{n-1}\lambda + a_n = 0$ have negative real parts. Then, all solutions of the scalar equation $y^{(n)} + a_1 y^{(n-1)} + \ldots + a_n y = 0$ are linear combinations of functions of the form $p_k(t)e^{\lambda t}$, where λ is a root of the characteristic polynomial and $p_k(t)$ is a polynomial in t of degree less than n. Because the exponentials $e^{\lambda t}$ all decay to zero as $t \to \infty$, so does every solution.

3 (a) Our assertion of asymptotic stability is equivalent to the characteristic polynomial $\lambda^2 + a_1\lambda + a_0 = 0$ having all its roots with negative real parts. If $a_1 > 0$ and $a_0 > 0$, the roots $\lambda_{1,2} = \frac{1}{2}(-a_1 \pm \sqrt{a_1^2 - 4a_0})$ will be either complex conjugate with negative real parts $-\frac{a_1}{2}$ or both real and negative. On the other hand, if $\lambda_{1,2}$ are complex conjugates with negative real parts, we will need $-\frac{a_1}{2} < 0$ and $4a_0 > a_1^2 > 0$. If a_0 and a_1 are both negative, $-a_1 \pm \sqrt{a_1^2 - 4a_0} < 0$ requires both $a_1 > 0$ and $a_0 > 0$.

(b) Again, we need to show that all roots of the characteristic polynomial $\lambda^3 + a_2\lambda^2 + a_1\lambda + a_0 = 0$ lie in the left half plane. Since one root

λ_1 is real, we have $\lambda^3 + a_2\lambda^2 + a_1\lambda + a_0 = (\lambda^2 + b_1\lambda + b_0)(\lambda - \lambda_1)$ with $a_2 = b_1 - \lambda_1$, $a_1 = b_0 - b_1\lambda_1$, and $a_0 = -b_0\lambda_1$, so $a_2 a_1 - a_0 = b_1(a_1 + \lambda_1^2)$. Suppose $a_2 > 0$, $a_1 > 0$, $a_0 > 0$, and $a_2 a_1 > a_0$. Then, $b_1 > 0$ and $b_0\lambda_1 < 0$. Since $b_0 > b_1\lambda_1$, we cannot have $b_0 < 0$ and $\lambda_1 > 0$. Instead, $b_0 > 0$ and $\lambda_1 < 0$. Thus, (a) implies that $\Re\lambda_{2,3} < 0$, so we have asymptotic stability. On the other hand, suppose $\lambda_1 < 0$ and $\Re\lambda_{2,3} < 0$. Then (a) implies that $b_0 > 0$ and $b_1 > 0$, so $a_2 = b_1 - \lambda_1 > 0$, $a_1 = b_0 - b_1\lambda_1 > 0$, $a_0 = -b_0\lambda_1 > 0$, and $a_2 a_1 - a_0 = b_1(a_1 + \lambda_1^2) > 0$.

We note that algebraic methods, as used in this exercise to guarantee stability, generalize to higher-order linear equations as the *Routh-Hurwitz criteria*. They are of practical importance in many fields of application.

4 (a) If the eigenvalues υ of B all have negative real parts, the entries of the matrix exponential e^{Bt} are all linear combinations of functions of the form $p(t)e^{\upsilon t}$ for polynomials p, so they decay exponentially as $t \to \infty$.

(b) Let $x_1(t) = \int_{-\infty}^{t} e^{B(t-s)} g_1(s)ds$. The integral will be well defined since $e^{B(t-s)}$ decays exponentially as $t \to \infty$. Furthermore, $\dot{x}_1 = Bx_1 + g_1(t)$. Analogously, $x_2(t) = -\int_t^{\infty} e^{C(t-s)} g_2(s)ds$ will be well defined for all t provided $-C$ is a stable matrix, and x_2 will then satisfy $\dot{x}_2 = Cx_2 + g_2$. We say that the homogeneous system has an *exponential dichotomy* under these hypotheses (cf. Coppel (1965)).

5 (a) The system $\dot{x}_1 = -x_1$, $\dot{x}_2 = -2x_2$ has the solution $x_1(t) = e^{-t}x_1(0)$, $x_2(t) = e^{-2t}x_2(0)$. Thus, the auxiliary condition $2\int_0^{\infty}(x_1^2(s) + 2x_2^2(s))\,ds = x_1^2(0) + x_2^2(0) = 1$ restricts the initial vector to the unit circle of the phase plane.

(b) If $x(t) = e^{-Pt}x(0)$, then the integral $2\int_0^{\infty} x^T(s)Px(s)ds = 2x^T(0)\left(\int_0^{\infty} e^{-Ps}Pe^{-Ps}ds\right)x(0)$ for $P = P^T$. However, $e^{-2Ps}P = \frac{d}{ds}\left(-\frac{1}{2}e^{-2Ps}\right)$ implies that $\int_0^{\infty} e^{-2Ps}Pds = -\frac{1}{2}e^{-2Ps}\Big|_0^{\infty} = \frac{1}{2}I$, since $-P$ is stable. Thus, $2\int_0^{\infty} x^T(s)Px(s)ds = 1$ restricts $x(0)$ to the unit sphere where $x^T(0)x(0) = 1$.

6 If $\sigma(x) = (y-z)^2$, then $\sigma'(x) = 2(y-z)(y'-z')$. Thus, if $y' = f(y, x)$, $z' = f(z, x)$, and $|f(y, x) - f(z, x)| \leq L|y - z|$ for some $L > 0$, $\sigma'(x) \leq 2L(y-z)^2 = 2L\sigma(x)$. Integrating $\frac{\sigma'(x)}{\sigma(x)} \leq 2L$ on $x \geq 0$ implies that $\sigma(x) \leq e^{2Lx}\sigma(0)$.

7 (a) The planar system $x' = (2-y)x$, $y' = (x-2)y$ has rest points at $\begin{pmatrix} x \\ y \end{pmatrix} = \begin{pmatrix} 0 \\ 0 \end{pmatrix}$ and $\begin{pmatrix} 2 \\ 2 \end{pmatrix}$ only.

(b) If $x(t) > 2$ and $y(t) > 2$, we will have $x' < 0$ and $y' > 0$, so $y(t) \geq y(0)$ and $x' \leq (2 - y(0))x$ imply that $x(t) \leq e^{[2-y(0)]t}x(0)$ as t increases, until $x(t) = 2$.

(c) The equation $\frac{dy}{dx} = \frac{(x-2)y}{(2-y)x}$ is separable and it implies that $H(x, y) = x - 2\ln x + y - 2\ln y = H(x(0), y(0))$ is constant on trajectories in the first quadrant.

(d) Consider the direction field about $\begin{pmatrix} 2 \\ 2 \end{pmatrix}$ and determine where x and y increase and decrease. This shows that trajectories with H constant are restricted to the first quadrant and that they encircle $\begin{pmatrix} 2 \\ 2 \end{pmatrix}$. Thus, H acts as a Liapunov function for this stable (but not asymptotically stable) rest point.

8 Since $V = pq$ is positive definite in the first quadrant where $\dot{V} = \dot{p}q + p\dot{q} = q^4 + p^2$ is also positive definite, the rest point is .

9 Here, $\dot{V} = z'B\dot{z} + \dot{z}'Bz = z'B(Az + H(z)) + (z'A' + H'(z))Bz = z'(BA + A'B)z + z'BH(z) + H'(z)Bz$ will be negative definite for z sufficiently small, so V is a Liapunov function.

10 We rewrite the equation as the equivalent system $\frac{dx}{dt} = y - F(x)$, $\frac{dy}{dt} = -g(x)$. Differentiating $V(x, y)$ along solutions implies that $dV/dt = g(x)(y - F(x)) + Y(-g(x)) = -g(x)F(x)$. Under our assumptions, $V \geq 0$ and $\dot{V} \leq 0$ near $x = 0$. In particular, $V > 0$ and $\dot{V} < 0$ in a punctured neighborhood of the origin if $g(x)F(x) > 0$, as when $xf(x) > 0$ and $xg(x) > 0$ near $x = 0$.

6
Singular perturbation methods

6.1 Regular and singular perturbations

Perturbation methods for ordinary differential equations use all the techniques we have learned and thus provide an excellent application for our acquired skills. They, in turn, are essential to current work in many areas of science, from fluid mechanics to biology and from control theory to quantum mechanics. In this final chapter, then, we will deemphasize the exercises and, instead, give a descriptive coverage of some singular perturbation methods, leaving extensive further analysis to O'Malley (1991) and other more advanced and specialized monographs listed in the references at the end of this chapter.

We will call a perturbation problem P_ϵ *regular* if its solutions y_ϵ feature smooth dependence on a small parameter ϵ. If P_ϵ is described through an equation $f(y, \epsilon) = 0$, we will require that $y_\epsilon \to Y_0$ as $\epsilon \to 0$, where Y_0 is defined by $f(Y_0, 0) = 0$, i.e., Y_0 satisfies P_0. Since ϵ usually represents a physically-meaningful parameter, letting ϵ tend to zero corresponds to neglecting the effect of such a small perturbation.

Example 1 (an algebraic equation) Consider parameter-dependent polynomials that maintain their degree near $\epsilon = 0$. Take, for example,

$$f(y, \epsilon) = y^2 - \epsilon y - 1 = 0.$$

The two solutions

$$\begin{aligned} y(\epsilon) &= \frac{1}{2}\left(-\epsilon \pm \sqrt{\epsilon^2 + 4}\right) \\ &= -\frac{\epsilon}{2} \pm \sqrt{1 + \frac{\epsilon^2}{4}} \end{aligned}$$

have the convergent power series expansions

$$y(\epsilon) = \pm 1 - \frac{\epsilon}{2} \pm \frac{\epsilon^2}{8} + 0\epsilon^3 + \dots$$

for any complex ϵ such that $|\epsilon| < 2$. The limits ± 1 obtained as $\epsilon \to 0$ are, indeed, defined by the limiting equation $Y_0^2 - 1 = 0$, so the perturbation problem is regular near $\epsilon = 0$. Moreover, each of the two solutions for P_0 generates a corresponding solution of P_ϵ for ϵ near zero.

We might, by contrast, consider the linear algebraic equation

$$\epsilon y + 1 = 0,$$

which is not meaningful when $\epsilon = 0$, and note that the usual power series approach for ϵ small cannot provide its solution. *Rescaling*, however, by setting

$$z = \epsilon y$$

converts the equation to

$$z + 1 = 0,$$

allowing us to define the solution

$$y = -\frac{1}{\epsilon}$$

for $\epsilon \neq 0$, corresponding to the root $z = -1$ (for all ϵ). There is clearly a big difference between the undefined problem obtained when $\epsilon = 0$ and the nearly unbounded solution found when ϵ is small. This is no regular perturbation!

Example 2 (an initial value problem) Consider the initial value problem

$$\dot{x} + \epsilon x = 0$$

for $x(0) = 1$ and $t \geq 0$. This linear differential equation has the corresponding auxiliary polynomial $\lambda + \epsilon = 0$ and the unique solution

$$x(t, \epsilon) = e^{-\epsilon t}$$

of the initial value problem has the convergent power series expansion

$$x(t, \epsilon) = 1 - \epsilon t + \frac{1}{2}\epsilon^2 t^2 + \dots$$

for any bounded t and any parameter ϵ. We, then seem to have a regular perturbation problem as $\epsilon \to 0$.

Note, however, that the limit of $x(t,\epsilon)$ as $t \to \infty$ is nonuniform because

$$x(t,\epsilon) \to \begin{cases} 1 & \text{for } t \text{ bounded as } \epsilon \to 0 \\ 0 & \text{for } t \gg 1/\epsilon \text{ as } \epsilon \to 0^+ \\ \infty & \text{for } t \gg -1/\epsilon \text{ as } \epsilon \to 0^-. \end{cases}$$

Here, the problem $P_\epsilon(x,t)$ is defined by $\{\dot{x} + \epsilon x = 0, x(0) = 1\}$ and its solution satisfies $\lim_{\epsilon\to 0} x(t,\epsilon) = X_0(t)$, for t bounded, where X_0 is the solution of $P_0(X_0,t)$: $\{\dot{X}_0 = 0,\ X_0(0) = 1\}$. To any order ϵ^j, we can obtain a unique Maclaurin series

$$x(t,\epsilon) = X_0(t) + \epsilon X_1(t) + \epsilon^2 X_2(t) + \ldots$$

about $\epsilon = 0$ with

$$X_j(t) \equiv \frac{1}{j!}\frac{d^j x(t,\epsilon)}{d\epsilon^j}\bigg|_{\epsilon=0} = t^j$$

for finite t. The regular problem, however, becomes *singular* (i.e., not regular) when $t \to \pm\infty$.

We shall henceforth follow the usual *convention* for singular perturbation problems by restricting the parameter ϵ to be real, *positive*, and asymptotically small (so $\epsilon \to 0^+$). We note that the breakdown of the regular perturbation method, when ϵt (in this example) becomes unbounded, is characteristic of many *secular* problems, such as those encountered by Poincaré and others in celestial mechanics at the end of the past century.

Example 3 The initial value problem

$$\epsilon\dot{x} = -x, \quad x(0) = 1$$

has the unique monotonic solution $x(t,\epsilon) = e^{-t/\epsilon}$ for $t \geq 0$, corresponding to the root $\lambda = -1/\epsilon$ of the corresponding characteristic (or auxiliary) polynomial. For $\epsilon < 0$, the solution would, instead, blow up for $t \ll -\epsilon$. For $\epsilon > 0$, however, $x(t,\epsilon) \to 0$ as $\epsilon \to 0$ for $t \gg \epsilon$ (and $x = 0$ is seen to be an asymptotically stable rest point).

Computationally, it is not simple to obtain an approximate solution, since the problem for $\epsilon > 0$ is *stiff* (i.e., stability requirements generally restrict the finite difference stepsize to being of order ϵ, even though the limiting solution to is nearly constant away from $t = 0$.) One often uses *upwinding* (i.e., backward difference approximations) to capture the rapidly decaying solution near $t = 0$ and to allow a reasonable stepsize to be used for $t > 0$. The resulting algorithms are popularly known as Gear's methods. Pictorially, we have

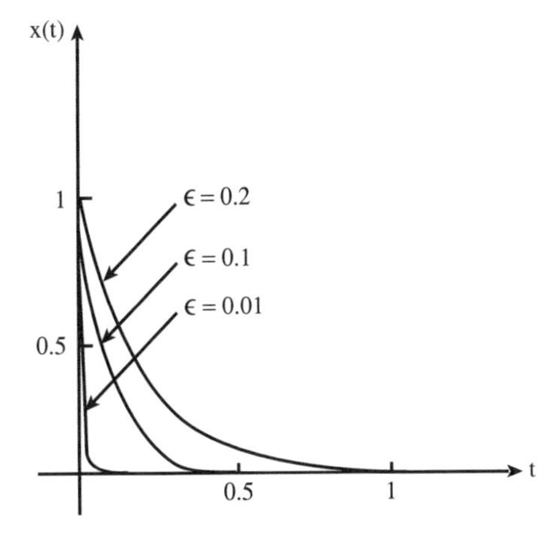

Figure 18. The steepening profiles $x(t, \epsilon)$ for $\epsilon = 0.2, 0.1$ and 0.01.

Note that the limiting solution

$$x(t, \epsilon) \to \begin{cases} 1, & t = 0 \\ 0, & t > 0 \end{cases}$$

is discontinuous as $\epsilon \to 0$ (in contrast to solutions of regular perturbation problems, which feature a uniform limit). Since $\dot{x}(0) = -\frac{1}{\epsilon}$, we should anticipate the rapid initial decrease in the solution. Uniqueness, however, requires x to remain positive for $t > 0$, so it is not surprising that the monotonically decreasing solution is soon nearly zero and quiescent.

The nonuniform convergence observed takes place in an *initial layer* of order ϵ thickness in t (to get $|x| < \delta$ for any small $\delta > 0$, we simply need $t > -\epsilon \ln \delta$). It is significant that the trivial *outer limit*

$$X_0(t) = 0$$

for any $t > 0$ satisfies the *reduced* (or *limiting*) *equation* $-X_0 = 0$ obtained by setting $\epsilon = 0$. We naturally lose the initial condition $x(0) = 1$ in the limit because the reduced equation has differential order zero, rather than one, and no initial value is needed to determine its solution. This explains why a discontinuous limit for x at $t = 0$ is necessary. (Such a cancellation of boundary conditions was examined more generally for boundary value problems in Wasow (1944).)

One can most conveniently describe the limiting solution to example 3 by introducing the stretched variable

$$\tau = t/\epsilon$$

(reflecting the size of the root of $\epsilon\lambda + 1 = 0$) and rewriting the problem as the parameter-free *inner problem*

$$\frac{dz}{d\tau} = -z, \quad z(0) = 1,$$

which has the unique solution

$$x(t, \epsilon) = z(\tau) = e^{-t/\epsilon} \quad \text{for } \tau \geq 0.$$

We note that the analogous use of stretched coordinate variables was basic to Prandtl's *boundary layer theory* in fluid mechanics and in other earlier work (cf. Goldstein (1969) and Van Dyke (1994)). Singular perturbations theory naturally inherited the terminology of inviscid fluid dynamics and fluid mechanics has continued to be its primary applications field.

Solutions to the vector initial value problem

$$\epsilon\dot{x} = Ax, \quad x(0) \text{ prescribed}$$

as $\epsilon \to 0$ can be found analogously whenever A is a constant stable matrix (whose eigenvalues lie strictly in the left half plane). The unique solution

$$x(t, \epsilon) = e^{At/\epsilon}x(0)$$

(expressed in terms of the matrix exponential) features an initial layer of nonuniform convergence near $t = 0$ with order ϵ thickness, as can be seen by again introducing the stretching $\tau = t/\epsilon$ and using the solution $z(\tau) = e^{A\tau}x(0)$ of the stretched problem $\frac{dz}{d\tau} = Az$, $z(0) = x(0)$, which decays exponentially to zero as $\tau \to \infty$. Computational techniques designed for such initial value problems with rapid change are presented, for example, in Hairer and Wanner (1991).

6.2 Linear initial value problems

Consider the nonhomogeneous vector (or matrix) equation

$$\epsilon\dot{x} = A(t)x + b(t) \tag{6.1}$$

on $t \geq 0$, where the square matrix $A(t)$ is smooth and stable, $b(t)$ is smooth, and $x(0)$ is prescribed. Note that no explicit solution is generally available, unless A is constant. Indeed, existence and uniqueness of the solution for $t > 0$ is in some doubt, because the Jacobian $\frac{1}{\epsilon}A(t)$ usually becomes undefined in the

$\epsilon \to 0$ limit. If we presume that $A(0)x(0) + b(0)$ is nonzero, then the large initial derivative requires rapid initial motion, (correctly) suggesting that we are faced with a singular perturbation problem with a thin initial layer where the solution converges nonuniformly.

We naturally first seek a formal outer solution

$$X(t, \epsilon) = X_0(t) + \epsilon X_1(t) + \epsilon^2 X_2(t) + \dots \tag{6.2}$$

of the system (6.1), using the regular perturbation procedure, expecting it to be appropriate (in some sense) away from $t = 0$. Since $\dot{X} = \dot{X}_0 + \epsilon \dot{X}_1 + \epsilon^2 \dot{X}_2 + \dots$, equating coefficients of successive powers of ϵ in the equation

$$\epsilon(\dot{X}_0 + \epsilon \dot{X}_1 + \dots) = A(t)(X_0 + \epsilon X_1 + \epsilon^2 X_2 + \dots) + b(t)$$

will imply

(from the ϵ^0 terms) : $AX_0 + b = 0$ or $X_0 = -A^{-1}b$
(from the ϵ^1 terms) : $AX_1 = \dot{X}_0$ or $X_1 = -A^{-1}\frac{d}{dt}(A^{-1}b) = A^{-2}(\dot{A}A^{-1}b - \dot{b})$
(from the ϵ^2 terms) : $AX_2 = \dot{X}_1$ or $X_2 = -A^{-1}\frac{d}{dt}(A^{-1}\frac{d}{dt}(A^{-1}b))$

etc. Thus, we uniquely obtain the *outer expansion*

$$X(t, \epsilon) = -A^{-1}b - \epsilon A^{-1}\frac{d}{dt}(A^{-1}b) - \epsilon^2 A^{-1}\frac{d}{dt}\left(A^{-1}\frac{d}{dt}(A^{-1}b)\right) + \dots \tag{6.3}$$

by formally proceeding termwise. We must certainly expect difficulties in attempting to use this expansion at turning points (where A becomes singular) or when either A or b loses smoothness. Expressions for successive coefficients $X_j = A^{-1}\frac{d}{dt}X_{j-1}$ might be generated using Maple, Mathematica, or some other software. However, we will generally approximate solutions using only a few terms of such an outer expansion, which we generally expect to diverge as $\epsilon \to 0$ (suggesting that using only a few terms is appropriate in practice, rather than slothful).

There is no reason to expect such an outer expansion X to match the imposed initial value $x(0)$, so we will now try to introduce the stretched variable

$$\tau = t/\epsilon$$

and to compensate for any initial discrepancy between $x(0)$ and $X(0, \epsilon)$ by asking that an initial layer *corrector*

$$z(\tau, \epsilon) \equiv x(t, \epsilon) - X(t, \epsilon) \tag{6.4}$$

satisfy the resulting homogeneous system

$$\frac{dz}{d\tau} = \epsilon(\dot{x} - \dot{X}) = A(\epsilon\tau)z, \tag{6.5}$$

with the initial value

$$z(0,\epsilon) = x(0) - X(0,\epsilon) \tag{6.6}$$

known in terms of the outer solution. Clearly, the outer expansion X will be asymptotically appropriate for $t > 0$ if $z(\tau,\epsilon) \to 0$ as $\tau = t/\epsilon \to \infty$. Moreover, $z(\tau,\epsilon)$ will be trivial for all $\tau \geq 0$ when $x(0) = X(0,\epsilon)$ and an initial layer is not needed. We will seek to solve the initial value problem for the *slowly varying* system (6.5) using the stability of the limiting coefficient matrix $A(0)$ and the assumed smoothness of A. Fortunately, Perron's stability theorem will imply the asymptotic stability of z as $\tau \to \infty$. Proceeding formally, we set

$$z(\tau,\epsilon) = z_0(\tau) + \epsilon z_1(\tau) + \epsilon^2 z_2(\tau) + \ldots, \tag{6.7}$$

asking that successive terms $z_j \to 0$ as $\tau \to \infty$. This presumes that the domain of attraction for the rest point $z = 0$ includes the initial value $z(0,\epsilon)$. Taking $\frac{dz}{d\tau} = \frac{dz_0}{d\tau} + \epsilon\frac{dz_1}{d\tau} + \epsilon^2\frac{dz_2}{d\tau} + \ldots$, we will use the Maclaurin series for $A(\epsilon\tau)$ and shall thereby need to satisfy

$$\frac{dz_0}{d\tau} + \epsilon\frac{dz_1}{d\tau} + \epsilon^2\frac{dz_2}{d\tau} + \ldots = \left[A(0) + \epsilon\tau\dot{A}(0) + \frac{1}{2}\epsilon^2\tau^2\ddot{A}(0) + \ldots\right]$$
$$\times\,(z_0 + \epsilon z_1 + \epsilon^2 z_2 + \ldots).$$

Equating coefficients successively, in both the differential equation and the initial condition, will require

$$\frac{dz_0}{d\tau} = A(0)z_0,\ \ z_0(0) = x(0) - X_0(0) = x(0) + A^{-1}(0)b(0),$$

$$\frac{dz_1}{d\tau} = A(0)z_1 + \tau\dot{A}(0)z_0, \quad z_1(0) = -X_1(0) = A^{-1}(0)\frac{d}{dt}(A^{-1}b)(0),$$

etc. Thus, we obtain the exponentially decaying leading term

$$z_0(\tau) = e^{A(0)\tau}(x(0) + A^{-1}(0)b(0)) \tag{6.8}$$

(invoking the stability of $A(0)$) and, then,

$$z_1(\tau) = e^{A(0)\tau}\left[A^{-1}(0)\frac{d}{dt}(A^{-1}b)(0) + \left(\int_0^\tau e^{-A(0)s}\dot{A}(0)e^{A(0)s}s\,ds\right)\right.$$
$$\left.\times\,(x(0) + A^{-1}(0)b(0))\right]. \tag{6.9}$$

We might anticipate that each

$$z_k(\tau) \equiv e^{A(0)\tau} b_k(\tau) \tag{6.10}$$

will decay to zero as $\tau \to \infty$, *roughly* like a polynomial (in τ) multiple of the matrix exponential $e^{A(0)\tau}$ (as in the scalar case, when the appropriate matrices commute and this statement is precise). Putting our formal results together, we have determined a *composite expansion*

$$x(t, \epsilon) = X(t, \epsilon) + z(t/\epsilon, \epsilon) \tag{6.11}$$

to the initial value problem for (6.1) to any desired number N of terms. Such a representation features a layer of nonuniform convergence of order ϵ thickness (where z remains asymptotically significant) near $t = 0$. Thus, we can readily generate any finite sequence

$$X^N(t, \epsilon) \equiv \sum_{j=0}^{N} (X_j(t) + Z_j(t/\epsilon))\epsilon^j \tag{6.12}$$

of approximations, for $N = 1, 2, 3, \ldots$. For any *fixed* $N \geq 0$, we can hope to bound the error; for instance, to show that

$$|x(t, \epsilon) - X^N(t, \epsilon)| \leq B_N \epsilon^{N+1}$$

as $\epsilon \to 0$, for some bounded constant B_N. (In more difficult applied problems, one most often relies on consistent analytical results and supporting computations rather than a rigorous estimate). Showing such a bound is equivalent to the assertion that the generated approximations X^N are (generalized) asymptotic expansions of the solution in the sense of Poincaré (cf. Erdélyi (1956)). We will henceforth use the convenient notation

$$x(t, \epsilon) \sim X(t, \epsilon) + z(t/\epsilon, \epsilon) \tag{6.13}$$

for such *asymptotic equality*.

For $N = 0$, we simply set

$$x(t, \epsilon) = X_0(t) + z_0(\tau) + u(t, \epsilon) \tag{6.14}$$

and show that $|u| \leq B\epsilon$ for some constant bound B. (The weaker conclusion that $u \to 0$ as $\epsilon \to 0$ is an often adequate alternative that is easier to verify.) Linearity implies that

$$\epsilon \frac{dx}{dt} = \epsilon \frac{dX_0}{dt} + \frac{dz_0}{d\tau} + \epsilon \frac{du}{dt} = A(t)(X_0 + z_0 + u) + b(t).$$

Using the definitions of X_0 and z_0, the remainder $u(t, \epsilon)$ must satisfy the nonhomogeneous initial value problem

$$\epsilon \frac{du}{dt} = A(t)u + (A(t) - A(0))z_0(\tau) + \epsilon \frac{d}{dt}(A^{-1}b), \; u(0, \epsilon) = 0. \tag{6.15}$$

We will estimate u by introducing the fundamental matrix $\mathcal{U}(t, \epsilon)$ for the corresponding homogeneous problem

$$\epsilon \frac{d\mathcal{U}}{dt} = A(t)\mathcal{U}, \;\; \mathcal{U}(0, \epsilon) = I, \tag{6.16}$$

noting that $\mathcal{U}(t, \epsilon) = e^{At/\epsilon}$ when A is constant and that $\mathcal{U}$ can, more generally, be found or approximated using successive approximations in the corresponding linear integral equation

$$\mathcal{U}(t, \epsilon) = I + \frac{1}{\epsilon} \int_0^t A(s)\mathcal{U}(s, \epsilon)ds. \tag{6.17}$$

Most critically, one can show the decay estimate

$$|\mathcal{U}(t, \epsilon)| \leq Me^{-\kappa t/\epsilon} \tag{6.18}$$

(in an appropriate norm) for positive constants M and κ, with $-\kappa$ often taken as the upper bound for the real parts of the eigenvalues of $A(0)$. Then, variation of parameters implies that the solution u of (6.15) is explicitly given by

$$u(t, \epsilon) = \mathcal{U}(t, \epsilon) \int_0^t \mathcal{U}^{-1}(s, \epsilon) \left[\frac{1}{\epsilon}(A(s) - A(0))z_0(s/\epsilon) + \frac{d}{ds}(A^{-1}b) \right] ds. \tag{6.19}$$

Due to the estimate (6.18), the calculation

$$e^{-\kappa t/\epsilon} \int_0^t e^{\kappa s/\epsilon} ds = \frac{\epsilon}{\kappa}(1 - e^{-\kappa t/\epsilon}) \leq \frac{\epsilon}{\kappa}$$

and the mean value theorem, $|u(t, \epsilon)| \leq \frac{\epsilon \widetilde{M}}{\kappa}$ for a bounded constant $\widetilde{M}$. In a similar fashion, one could estimate the order ϵ^{N+1} error $|x - X^N|$ for any integer $N > 0$. Because the *initial layer correction* $z(\tau, \epsilon)$ becomes asymptotically negligible as $\tau \to \infty$, the asymptotic solution for x will be provided by the outer expansion $X(t, \epsilon)$ above any fixed positive value of t.

For τ bounded, however, we get an *inner expansion* of the solution $x(t, \epsilon)$ by formally expanding the composite expansion

$$X(\epsilon\tau, \epsilon) + z(\tau, \epsilon)$$

(a function of τ) in its Maclaurin series about $\epsilon = 0$, i.e.,

$$\begin{aligned} X(\epsilon\tau, \epsilon) + z(\tau, \epsilon) &= X_0(\epsilon\tau) + \epsilon X_1(\epsilon\tau) + \epsilon^2 X_2(\epsilon\tau) \\ &\quad + \ldots + z_0(\tau) + \epsilon z_1(\tau) + \epsilon^2 z_2(\tau) + \ldots \\ &= (X_0(0) + \epsilon\tau \dot{X}_0(0) + \ldots) + \epsilon(X_1(0) + \ldots) + \ldots \\ &\quad + z_0(\tau) + \epsilon z_1(\tau) + \ldots \\ &= (z_0(\tau) + X_0(0)) + \epsilon(z_1(\tau) + \tau \dot{X}_0(0) + X_1(0)) + \ldots . \end{aligned} \tag{6.20}$$

This expansion is, more traditionally, found directly by seeking an *inner solution* $v(\tau, \epsilon)$, which, instead, satisfies the stretched initial value problem

$$\frac{dv}{d\tau} = A(\epsilon\tau)v + b(\epsilon\tau), \quad v(0) = x(0) \tag{6.21}$$

and which has an inner expansion

$$v(\tau, \epsilon) = v_0(\tau) + \epsilon v_1(\tau) + \epsilon^2 v_2(\tau) + \ldots \tag{6.22}$$

for finite values of τ that *matches* the known outer expansion $X(t, \epsilon)$ as $\tau \to \infty$ and $t \to 0$, so, e.g., $\lim_{\tau\to\infty} v_0(\tau) = \lim_{t\to 0} X_0(t)$. The terms v_j will typically grow algebraically with τ, in contrast to the exponentially decaying z_js. Although the expansion will, in general, diverge, it has substantial practical importance.

Matching rules for inner and outer expansions and the construction of related uniformly valid composite expansions were the principal impact of the work of Van Dyke (1964, 1975), whereas Cole (1968), Eckhaus (1979), and Kevorkian and Cole (1996) rely on more careful and intricate limit process and multiple scale expansions. Our approach of correcting outer expansions in regions of nonuniform convergence is developed in O'Malley (1991), Lions (1973), and Vasil'eva et al. (1995).

6.3 Nonlinear initial value problems

Consider the vector system

$$\begin{cases} \dot{x} = f(x, y, t, \epsilon) \\ \epsilon\dot{y} = g(x, y, t, \epsilon) \end{cases} \tag{6.23}$$

on the interval $t \geq 0$, assuming that f and g are smooth m and n dimensional vectors, that $x(0)$ and $y(0)$ are prescribed vectors, that the $n \times n$ Jacobian matrix g_y remains stable everywhere, and that the stretched (inner) problem

$$\frac{dz}{ds} = g(x(0), z, 0, 0) \tag{6.24}$$

has an asymptotically stable solution tending to a rest point $\varphi(x(0))$ as $s \to \infty$ for all initial vectors $z(0)$. A special case of (6.23) is, of course, the linear scalar problem of section 6.2.

We shall construct an asymptotic solution of (6.23) in the form

$$\begin{cases} x(t,\epsilon) = X(t,\epsilon) + \epsilon\xi(\tau,\epsilon) \\ y(t,\epsilon) = Y(t,\epsilon) + \eta(\tau,\epsilon), \end{cases} \tag{6.25}$$

where all terms have power series expansions in ϵ and the initial layer correction $\binom{\epsilon\xi}{\eta}$ tends to zero as the stretched variable

$$\tau = t/\epsilon$$

tends to infinity. Thus, the limiting solution for $t > 0$ will be provided by the outer solution

$$\begin{pmatrix} X(t,\epsilon) \\ Y(t,\epsilon) \end{pmatrix}.$$

Moreover, the *slow* variable x will converge uniformly to $X_0(t) = X(t,0)$ as $\epsilon \to 0$, while the *fast* variable y will converge nonuniformly near $t = 0$. Higher derivatives of the solution will agree with the results of formal differentiations, so they will generally blow up algebraically within the initial layer as $\epsilon \to 0$.

A meaningful application of our theory in enzyme kinetics is presented in Segel and Slemrod (1989) (cf. also Murray (1989)). It involves the two-dimensional initial value problem

$$\begin{cases} \frac{dx}{dt} = -x + (x + \mu - \lambda)y, & x(0) = 1 \\ \epsilon\frac{dy}{dt} = x - (x + \mu)y, & y(0) = 0. \end{cases} \tag{6.26}$$

The physical interpretations of the small parameter ϵ are quite significant, in addition to the essential role ϵ plays in determining the thickness of the initial t-interval of rapid change for y and for derivatives of the solution.

Formally, the outer expansion

$$\begin{pmatrix} X(t,\epsilon) \\ Y(t,\epsilon) \end{pmatrix} \sim \sum_{k=0}^{\infty} \begin{pmatrix} X_k(t) \\ Y_k(t) \end{pmatrix} \epsilon^k \tag{6.27}$$

must satisfy the system termwise for each fixed $t > 0$, since the initial layer correction $\binom{\epsilon\xi}{\eta}$ is asymptotically negligible there. Expanding the right-hand side of (6.23) in a Maclaurin series about $(X_0, Y_0, t, 0)$, we find that the leading coefficient $\binom{X_0}{Y_0}$ must satisfy the nonlinear differential-algebraic system

$$\begin{cases} \dot{X}_0 = f(X_0, Y_0, t, 0) \\ 0 = g(X_0, Y_0, t, 0) \end{cases} \tag{6.28}$$

together with the initial value $X_0(0) = x(0)$ for the slow vector x. Higher-order terms $\binom{X_j}{Y_j}$ must satisfy the linearized system

$$\begin{cases} \dot{X}_j = f_x(X_0, Y_0, t, 0)X_j + f_y(X_0, Y_0, t, 0)Y_j + \alpha_{j-1}(t) \\ 0 = g_x(X_0, Y_0, t, 0)X_j + g_y(X_0, Y_0, t, 0)Y_j + \beta_{j-1}(t), \end{cases} \tag{6.29}$$

where the forcing vector $\binom{\alpha_{j-1}(t)}{\beta_{j-1}(t)}$ is completely determined by preceding terms $\binom{X_k(t)}{Y_k(t)}$ for $k < j$ and their derivatives.

Since g_y is nonsingular, the implicit function theorem implies that we can solve the "algebraic" equation $g(X_0, Y_0, t, 0) = 0$ in a locally unique way to define

$$Y_0 = \varphi(X_0, t). \tag{6.30}$$

More explicitly, differentiation of the constraint ($g = 0$) with respect to t implies that

$$g_x(X_0, Y_0, t, 0)f(X_0, Y_0, t, 0) + g_y(X_0, Y_0, t, 0)\dot{Y}_0 + g_t(X_0, Y_0, t, 0) = 0,$$

so if we let $Y_0(0) = \varphi(x(0), 0)$ be any *isolated* root of $g(x(0), Y_0(0), 0, 0) = 0$, we will determine a corresponding solution of (6.28) on some interval $t > 0$ by numerically integrating the nonlinear initial value problem

$$\begin{cases} \dot{X}_0 = f(X_0, Y_0, t, 0) \\ \dot{Y}_0 = -g_y^{-1}(X_0, Y_0, t, 0)[g_x(X_0, Y_0, t, 0)f(X_0, Y_0, t, 0) + g_t(X_0, Y_0, t, 0)] \end{cases}$$

forward in time, using the initial vector $\binom{x(0)}{Y_0(0)}$. Care must be taken to ensure that the computed solution does not drift off the constraint. Alternatively, if we explicitly know the root $Y_0 = \varphi(X_0, t)$, solving the lower-dimensional initial value problem

$$\dot{X}_0 = f(X_0, \varphi(X_0, t), t, 0), \;\; X_0(0) = x(0) \tag{6.31}$$

has obvious advantages over integrating the higher-dimensional system for $\binom{X_0}{Y_0}$. Assuming $\binom{X_0(t)}{Y_0(t)}$ is known on the interval $0 \le t \le T$, for some $T > 0$, (6.29) implies that we will successively have

$$Y_j(t) = -g_y^{-1}(X_0, Y_0, t, 0)[g_x(X_0, Y_0, t, 0)X_j + \beta_{j-1}(t)], \tag{6.32}$$

and there simply remains a linear system for X_j in the form

$$\dot{X}_j = \frac{d}{dx}[f(X_0, \varphi(X_0, t), t, 0)]X_j + \gamma_{j-1}(t)$$

with γ_{j-1} known successively. Moreover, since differentiation of (6.31) with respect to the initial vector $x(0)$ implies that the corresponding homogeneous system $\dot{z} = \frac{d}{dx}[f(X_0, \varphi(X_0, t), t, 0)]z$ has the fundamental matrix $\frac{\partial X_0(t)}{\partial x(0)}$, variation of parameters provides the explicit solution

$$X_j(t) = \frac{\partial X_0(t)}{\partial x(0)} \left[X_j(0) + \int_0^t \left(\frac{\partial X_0(s)}{\partial x(0)} \right)^{-1} \gamma_{j-1}(s)ds \right] \tag{6.33}$$

for X_j and, thereby determines Y_j. Thus, we can readily generate an outer expansion $\binom{X(t,\epsilon)}{Y(t,\epsilon)}$ corresponding to any initial vectors $X(0, \epsilon)$ and $Y_0(0)$. Note that the initial vector $X(0, \epsilon) = x(0) - \epsilon\xi(0, \epsilon)$ depends on the undetermined initial layer correction term ξ.

Since $\dot{x} = \frac{dX}{dt} + \frac{d\xi}{d\tau} = f(x, y, t, \epsilon)$ and $\epsilon\dot{y} = \epsilon\frac{dY}{dt} + \frac{d\eta}{d\tau} = g(x, y, t, \epsilon)$, while $\frac{dX}{dt} = f(X, Y, t, \epsilon)$ and $\epsilon\frac{dY}{dt} = g(X, Y, t, \epsilon)$, the representations (6.25) for x and y imply that the initial layer correction $\binom{\epsilon\xi}{\eta}$ must satisfy the nonlinear system

$$\begin{cases} \frac{d\xi}{d\tau} = f(X + \epsilon\xi, Y + \eta, \epsilon\tau, \epsilon) - f(X, Y, \epsilon\tau, \epsilon) \\ \frac{d\eta}{d\tau} = g(X + \epsilon\xi, Y + \eta, \epsilon\tau, \epsilon) - g(X, Y, \epsilon\tau, \epsilon) \end{cases} \tag{6.34}$$

on the interval $\tau \geq 0$, with all functions of t being evaluated at $\epsilon\tau$. Moreover, $\binom{\xi}{\eta}$ must decay to zero as $\tau \to \infty$, and the initial condition for y directly implies the initial condition

$$\eta(0, \epsilon) = y(0) - Y(0, \epsilon) \tag{6.35}$$

for η, which varies with the undetermined outer expansion vector Y. Setting $\epsilon = 0$ in (6.34) and (6.35) shows that $\binom{\xi_0}{\eta_0}$ must satisfy the limiting nonlinear system

$$\begin{cases} \frac{d\xi_0}{d\tau} = f(X_0(0), Y_0(0) + \eta_0, 0, 0) - f(X_0(0), Y_0(0), 0, 0) \\ \frac{d\eta_0}{d\tau} = g(X_0(0), Y_0(0) + \eta_0, 0, 0) - g(X_0(0), Y_0(0), 0, 0), \end{cases}$$

with $\eta_0(0) = y(0) - Y_0(0)$. Because $X_0(0) = x(0)$, $g(x(0), Y_0(0), 0, 0) = 0$, and $Y_0(0) = \varphi(x(0), 0)$, it follows that η_0 must satisfy the nonlinear initial value problem

$$\frac{d\eta_0}{d\tau} = g[x(0), \varphi(x(0), 0) + \eta_0, 0, 0], \; \eta_0(0) = y(0) - \varphi(x(0), 0) \tag{6.36}$$

on $\tau \geq 0$. Note that (6.36) coincides with the stretched problem (6.24) with the inner variable $z \equiv \eta_0 + \varphi(x(0), 0)$. By our stability assumption, we are guaranteed a unique exponentially decaying solution $\eta_0(\tau)$ for all $\tau \geq 0$. [In particular, $\eta_0(\tau) \equiv 0$ if $y(0) = \varphi(x(0), 0)$]. Knowing $\eta_0(\tau)$, existentially or through a numerical computation, uniquely determines the vector

$$\xi_0(\tau) = \int_\tau^\infty [f(x(0), Y_0(0), 0, 0) - f(x(0), Y_0(0) + \eta_0(s), 0, 0)]ds, \quad (6.37)$$

which necessarily also decays exponentially to zero as $\tau \to \infty$. In the process, note that we have specified the initial vector

$$X_1(0) = -\xi_0(0) \quad (6.38)$$

for the first-order outer term X_1 of the slow vector x, which was needed to completely specify the full first-order term $\binom{X_1(t)}{Y_1(t)}$ of the outer expansion. Equation (6.38) follows since the ϵ-coefficients in the initial condition $x(0) = X(0, \epsilon) + \epsilon\xi(0, \epsilon)$ must balance.

Proceeding termwise in (6.34), we successively obtain linear systems of the form

$$\begin{cases} \frac{d\xi_j}{d\tau} = f_y(x(0), Y_0(0) + \eta_0(\tau), 0, 0)\eta_j(\tau) + \gamma_{j-1}(\tau) \\ \frac{d\eta_j}{d\tau} = g_y(x(0), Y_0(0) + \eta_0(\tau), 0, 0)\eta_j(\tau) + \delta_{j-1}(\tau) \end{cases}$$

for each $j > 0$, where the forcing term $\binom{\gamma_{j-1}(\tau)}{\delta_{j-1}(\tau)}$ is known successively in terms of the earlier exponentially decaying coefficients $\binom{\xi_k(\tau)}{\eta_k(\tau)}$ (for $k < j$) and their derivatives. The differential equation (6.36) for $\eta_0(\tau)$ implies that

$$\frac{d}{d\tau}\left(\frac{d\eta_0}{d\eta_0(0)}\right) = g_y(x(0), Y_0(0) + \eta_0, 0, 0)\frac{d\eta_0}{d\eta_0(0)},$$

i.e., $\frac{d\eta_0}{d\eta_0(0)}$ is a fundamental matrix for the homogeneous system corresponding to that for η_j. Variation of parameters thereby provides the unique decaying solution

$$\eta_j(\tau) = \frac{d\eta_0(\tau)}{d\eta_0(0)}\left[-Y_j(0) - \int_\tau^\infty \left(\frac{d\eta_0(s)}{d\eta_0(0)}\right)^{-1} \gamma_{j-1}(s)ds\right], \quad (6.39)$$

since the ϵ^j coefficient in the initial condition for y requires $Y_j(0) + \eta_j(0) = 0$. [Note that $\frac{d\eta_0(\tau)}{d\eta_0(0)} = e^{A(0)\tau}$ for linear problems with $g_y(x(0), Y_0(0) + \eta_0, 0, 0) \equiv A(0)$.] Knowing η_j, in turn, uniquely determines the exponentially decaying vector

$$\xi_j(\tau) = -\int_\tau^\infty [f_y(x(0), Y_0(0) + \eta_0(s), 0, 0)\eta_j(s) + \gamma_{j-1}(s)]ds. \quad (6.40)$$

Moreover, the ϵ^{j+1} coefficient in the initial condition for x then specifies the initial value

$$X_{j+1}(0) = -\xi_j(0)$$

needed to completely specify the $(j+1)$-st term $\binom{X_{j+1}}{Y_{j+1}}$ in the outer expansion. Thus, we have shown that a formal asymptotic solution of (6.23) in the form (6.25) can be constructed termwise to any order, without difficulty.

The only possibility of taking a misstep in the procedure outlined occurs when we select the root $Y_0(0)$ of the limiting algebraic equation $g(x(0), Y_0(0), 0, 0) = 0$. When there are several roots, the solution $\eta_0(\tau)$ of the resulting stability problem (6.36) might converge to a steady-state $\tilde{Y}_0(0) - Y_0(0)$ different than zero, suggesting that another root $\tilde{Y}_0(0)$, instead of $Y_0(0)$, should have been initially selected. Such an apparent mismatch indicates that the asymptotic solution is unique, with the appropriate root $Y_0(0)$ to use being determined by the prescribed initial value $y(0)$. One can simply demonstrate the situation by considering the scalar example $\epsilon \dot{y} = y(1-y^2)$, which has rest points $y = 0$ and ± 1.

Asymptotic convergence could be proved (following ideas independently developed about 1950 by Tikhonov at Moscow State and Levinson at MIT). One shows that, for every $N > 1$, there is a unique solution

$$\begin{pmatrix} x(t,\epsilon) \\ y(t,\epsilon) \end{pmatrix}$$

satisfying

$$\left| x(t,\epsilon) - X_0(t) - \sum_{j=1}^{N-1} (X_j(t) + \xi_{j-1}(t/\epsilon))\epsilon^j \right| \leq K_N \epsilon^N$$

and

$$\left| y(t,\epsilon) - \sum_{j=0}^{N-1} (Y_j(t) + \eta_j(t/\epsilon))\epsilon^j \right| \leq L_N \epsilon^N$$

in the uniform norm, for bounded constants K_N and L_N and ϵ sufficiently small (cf. Smith (1985) and Vasil'eva et al. (1995)). Because the terms of

$$\begin{pmatrix} \xi \\ \epsilon\eta \end{pmatrix} \to 0$$

as $\tau \to \infty$, the solution constructed will coincide asymptotically with the outer expansion

$$\begin{pmatrix} X(t,\epsilon) \\ Y(t,\epsilon) \end{pmatrix}$$

for each fixed $t > 0$. The fast variable y, however, will feature a thin initial layer of nonuniform convergence near $t = 0$, as will derivatives of both x and y (which will generally be algebraically unbounded with respect to ϵ at $t = 0$).

6.4 Two-point problems

Example 4 Let us first consider the scalar equation

$$\epsilon \ddot{x} + \dot{x} = 0 \tag{6.41}$$

on $0 \leq t \leq 1$ with the prescribed endvalues $x(0) = 1$ and $x(1) = 2$. Because the differential equation has the linearly independent solutions 1 and $e^{-t/\epsilon}$, any solution must have the form

$$x(t, \epsilon) = c_1 + c_2 e^{-t/\epsilon}$$

for ϵ-dependent coefficients c_j, independent of t. Applying the boundary conditions determines the c_js and the unique solution

$$x(t, \epsilon) = (2 - e^{-1/\epsilon} - e^{-t/\epsilon})/(1 - e^{-1/\epsilon}). \tag{6.42}$$

Neglecting the asymptotically small terms $e^{-1/\epsilon}$ provides the uniformly valid estimate

$$|x(t, \epsilon) - 2 + e^{-t/\epsilon}| \leq \epsilon^N$$

throughout $0 \leq t \leq 1$ for any integer $N > 0$, provided ϵ is sufficiently small. Thus, the solution obtained has the asymptotic form

$$x(t, \epsilon) \sim X(t, \epsilon) + \xi(\tau, \epsilon) \tag{6.43}$$

for the outer solution $X(t, \epsilon) \equiv 2$ and the initial layer correction $\xi(\tau, \epsilon) \equiv -e^{-\tau}$ expressed in terms of the stretched variable $\tau = t/\epsilon$. Pictorially, the solution is sketched in Figure 19.

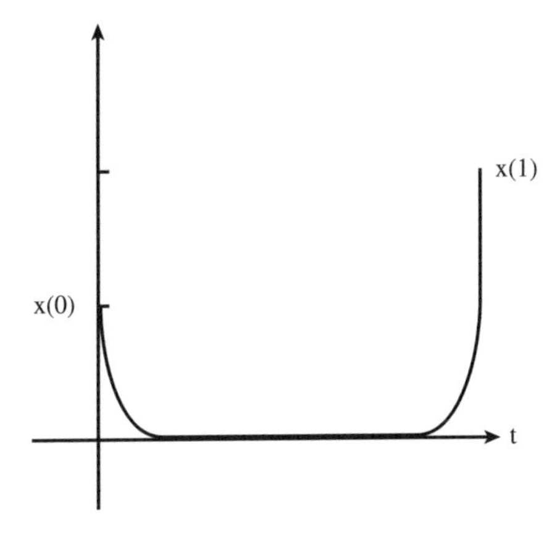

Figure 19. The solution of $\epsilon \ddot{x} + \dot{x} = 0$, $x(0) = 1$, $x(1) = 2$ for ϵ small.

Note that the outer solution satisfies the reduced (limiting) problem $\dot{X} = 0$, $X(1) = 2$. It is, asymptotically, only incorrect near $t = 0$, where the initial condition $x(0) = 1$ is violated. Thus, the added initial layer correction $\xi(t/\epsilon, \epsilon)$ provides the needed nonuniform convergence in a thin layer there.

Example 5 The linear variable-coefficient scalar equation

$$\epsilon \ddot{x} + a(t)\dot{x} + b(t)x = 0, \tag{6.44}$$

with prescribed bounded endvalues $x(0)$ and $x(1)$, also has an asymptotic solution of the form

$$x(t, \epsilon) \sim X(t, \epsilon) + \xi(\tau, \epsilon), \tag{6.45}$$

where the outer solution $X(t, \epsilon)$ and the initial layer correction $\xi(\tau, \epsilon)$ both have power series expansions in ϵ, with $\xi \to 0$ as the stretched initial layer coordinate $\tau = t/\epsilon \to \infty$, provided $a(t)$ and $b(t)$ are smooth and $a(t) > 0$ throughout $[0, 1]$. This conclusion is far from obvious, however.

It follows formally that the outer expansion

$$X(t, \epsilon) \sim \sum_{j=0}^{\infty} X_j(t)\epsilon^j$$

must satisfy the differential equation as a power series in ϵ, as well as satisfying the terminal condition. Equating coefficients successively, in the usual manner, requires

$$a(t)\dot{X}_0 + b(t)X_0 = 0, \;\; X_0(1) = x(1),$$
$$a(t)\dot{X}_1 + b(t)X_1 = -\ddot{X}_0, \; X_1(1) = 0,$$

etc. Thus,

$$X_0(t) = e^{\int_t^1 \frac{b(s)}{a(s)} ds} x(1),$$

$$X_1(t) = e^{\int_t^1 \frac{b(s)}{a(s)} ds} \int_t^1 \frac{1}{a(r)} \left[\left(\frac{b(r)}{a(r)}\right)^2 - \frac{d}{dr}\left(\frac{b(r)}{a(r)}\right)\right] dr,$$

etc. We note that complications would occur if the coefficient $a(t)$ had a zero within $[0, 1]$.

Likewise, the initial layer correction $\xi(\tau, \epsilon)$ must satisfy the stretched homogeneous equation

$$\frac{1}{\epsilon}\frac{d^2\xi}{d\tau^2} + \frac{1}{\epsilon}a(\epsilon\tau)\frac{d\xi}{d\tau} + b(\epsilon\tau)\xi = 0$$

as a power series $\xi(\tau, \epsilon) \sim \sum_{j=0}^{\infty} \xi_j(\tau)\epsilon^j$ in ϵ for $\tau \geq 0$, together with the initial condition $\xi(0, \epsilon) \sim x(0) - X(0, \epsilon)$ and the matching condition that $\xi \to 0$ as $\tau \to \infty$. Using the Maclaurin series for $a(\epsilon\tau)$ and $b(\epsilon\tau)$, we equate coefficients formally and naturally ask that

$$\frac{d^2\xi_0}{d\tau^2} + a(0)\frac{d\xi_0}{d\tau} = 0, \quad \xi_0(0) = x(0) - X_0(0), \quad \xi_0 \to 0 \text{ as } \tau \to \infty,$$

$$\frac{d^2\xi_1}{d\tau^2} + a(0)\frac{d\xi_1}{d\tau} + \tau\dot{a}(0)\frac{d\xi_0}{d\tau} + b(0)\xi_0 = 0, \ \xi_1(0) = -X_1(0), \ \xi_1 \to 0 \text{ as } \tau \to \infty$$

etc. Thus, we obtain the decaying terms

$$\xi_0(\tau) = e^{-a(0)\tau}(x(0) - X_0(0))$$

and

$$\xi_1(\tau) = e^{-a(0)\tau}\left[-X_1(0) - \frac{(\dot{a}(0) + b(0))}{a(0)}\tau - \frac{1}{2}\dot{a}(0)\tau^2\right],$$

etc. Writing only the lowest-order terms, we have the uniformly valid representation

$$x(t, \epsilon) \sim \left[e^{\int_t^1 \frac{b(s)}{a(s)}ds}x(1) + e^{-a(0)t/\epsilon}\left[x(0) - e^{\int_0^1 \frac{b(s)}{a(s)}ds}x(1)\right]\right] + \epsilon(\ldots). \quad (6.46)$$

We note that $\dot{x}$ and all higher derivatives will generally become unbounded at $t = 0$ when $\epsilon \to 0$. Analogously, if we instead had $a(t) < 0$ throughout the interval, we would need a terminal layer of nonuniform convergence near $t = 1$, which would be asymptotically described in terms of the stretched variable $\sigma = (1 - t)/\epsilon$. The turning point situation, when $a(t)$ has zeros within the interval, is much more complicated.

Example 6 Now consider the linear two-point problem

$$\epsilon\ddot{x} + t\dot{x} = 0 \text{ on } -1 \leq t \leq 1 \quad (6.47)$$

with a *turning point* at $t = 0$ and $x(\pm 1)$ given. Since 1 and $\int^t e^{-s^2/2\epsilon}$ are linearly independent solutions of this second-order homogeneous equation, its general solution must be of the form

$$x(t, \epsilon) = k + \left(\int_{-1}^{t} e^{-s^2/2\epsilon}ds\right)c$$

for (ϵ-dependent) constants k and c. Applying the boundary conditions implies the unique solution

$$x(t,\epsilon) = x(-1) + \left[\frac{\int_{-1}^{t} e^{-s^2/2\epsilon} ds}{\int_{-1}^{1} e^{-s^2/2\epsilon} ds} \right] (x(1) - x(-1)). \tag{6.48}$$

We note that the integrands are both even and asymptotically negligible, except for s near zero. Thus, the bracketed ratio of integrals is asymptotic to $+1$ for any fixed $t > 0$, to 0 for any $t < 0$, and to $\frac{1}{2}$ for $t = 0$. This implies that

$$x(t,\epsilon) \sim \begin{cases} x(-1) & \text{for } t < 0 \\ \frac{1}{2}(x(1) + x(-1)) & \text{for } t = 0 \\ x(1) & \text{for } t > 0. \end{cases} \tag{6.49}$$

The constant limiting solution $X_0(t)$ away from the turning point $t = 0$ clearly satisfies the *reduced* problem

$$\begin{cases} t\dot{X}_0 = 0,\ X_0(1) = x(1) & \text{for } t > 0 \\ \text{and} & \\ t\dot{X}_0 = 0,\ X_0(-1) = x(-1) & \text{for } t < 0. \end{cases}$$

The solution of the boundary value problem is clearly a function of the stretched variable $\xi \equiv t/\sqrt{\epsilon}$, and the solution's asymptotics as $\xi \to \pm\infty$ could be analyzed in terms of that variable or the special (error) function

$$\text{erf}\left(\frac{t}{2\epsilon}\right) \equiv \frac{2}{\sqrt{\pi}} \int_0^{t/\sqrt{2\epsilon}} e^{-s^2} ds.$$

In particular, this shows how the rapid transition from $x(-1)$ to $x(1)$ occurs in an order $\sqrt{\epsilon}$ neighborhood of $t = 0$.

Note that our results for the linear variable-coefficient scalar equation (6.44) with coefficient $a(t) > 0$ on $0 < t \leq 1$ suggest that we now have a layer at $t = 0^+$, whereas those for $a(t) < 0$ on $-1 \leq t < 0$ suggest a layer at $t = 0^-$. A difference here is that the order $\sqrt{\epsilon}$ layer of nonuniform convergence is now thicker than the order ϵ layers encountered when a had a fixed sign locally. This difference would be readily confirmed if numerical techniques were tried to obtain an approximate solution.

Example 7 If we consider the previous equation with a simple change in sign, i.e.,

$$\epsilon\ddot{x} - t\dot{x} = 0 \tag{6.50}$$

on $-1 \leq t \leq 1$, with $x(\pm 1)$ given, we should expect endpoint layers rather than a transition layer at the turning point. Indeed, any solution of the differential equation must have the form

$$x(t,\epsilon) = k + \left(\int_0^t e^{s^2/2\epsilon}\,ds\right) c,$$

so the boundary conditions imply the unique monotonic solution

$$x(t,\epsilon) = \frac{1}{2}(x(1) + x(-1)) + \frac{1}{2}\left[\frac{\int_0^t e^{s^2/2\epsilon}\,ds}{\int_0^1 e^{s^2/2\epsilon}\,ds}\right](x(1) - x(-1)). \tag{6.51}$$

Here, the ratio of integrals has the limit -1 as $\epsilon \to 0$ for $t = -1$, 0 for $-1 < t < 1$, and 1 for $t = 1$, since the major contribution to each integral comes from near the upper endpoint.

The limiting solution is therefore the average

$$\frac{1}{2}(x(1) + x(-1))$$

within $[-1, 1]$, and there is nonuniform convergence to the prescribed endvalue near both endpoints. We note that $\dot{x}$ has order $1/\epsilon$ at $t = \pm 1$, suggesting that the endpoint layers of nonuniform convergence now are of order ϵ thick. Nothing unusual happens at the turning point $t = 0$ where the limiting solution remains constant. A careful examination of the limiting behavior shows that $\int_0^t e^{s^2/2\epsilon}\,ds \sim \frac{\epsilon}{x} e^{x^2/2\epsilon}$ for $\frac{x^2}{\epsilon} \gg 1$. This implies the uniformly valid asymptotic solution

$$x(t,\epsilon) \sim \frac{1}{2}[x(1) + x(-1)] + \frac{1}{2}[x(-1) - x(1)]e^{-(t+1)/\epsilon} \tag{6.52}$$

$$+ \frac{1}{2}[x(1) - x(-1)]e^{(t-1)/\epsilon}$$

to all orders ϵ^N.

Example 8 The nonlinear problem consisting of *Burgers' equation*

$$\epsilon\ddot{x} = 2x\dot{x} \tag{6.53}$$

on $-1 \le t \le 1$, with prescribed endvalues $x(\pm 1)$, can nearly be solved exactly. (We note that this ordinary differential equation is the steady-state limit of Burgers' partial differential equation, encountered in turbulence modeling and elsewhere.) If we integrate the equation once, we get $\epsilon\dot{x} = x^2 - A^2$ for a real (and nonnegative) or a purely imaginary constant A. Then, separating variables, we get

$$x(t,\epsilon) = -A\tanh(A(t-\beta)/\epsilon) = -A\left[\frac{1 - e^{-A(t-\beta)/2\epsilon}}{1 + e^{-A(t-\beta)/2\epsilon}}\right]. \tag{6.54}$$

The boundary values thereby provide the two relations

$$e^{A\beta/2\epsilon} = \left(\frac{A+x(1)}{A-x(1)}\right) e^{A/2\epsilon} = \left(\frac{A+x(-1)}{A-x(-1)}\right) e^{-A/2\epsilon} \tag{6.55}$$

for $A(\epsilon)$ and $\beta(\epsilon)$. We will separately study several cases.

Let us begin our asymptotic treatment by considering the symmetric situation when, say, $x(1) = -x(-1) < 0$. Then, we will necessarily have the zero location $\beta = 0$ and the amplitude A must satisfy the transcendental equation

$$A - x(-1) = (A + x(-1))e^{-A/2\epsilon}.$$

Using iteration, we get the (unique) asymptotic solution

$$A = x(-1) + 2x(-1)e^{-x(-1)/2\epsilon} + \dots. \tag{6.56}$$

Because $x(-1) > 0$, omitting the asymptotically negligible term $e^{-x(-1)/2\epsilon}$ and further terms in A provides the asymptotic solution

$$x(t,\epsilon) \sim -x(-1)\left[\frac{1-e^{-x(-1)t/2\epsilon}}{1+e^{-x(-1)t/2\epsilon}}\right].$$

Moreover, we will then have

$$x(t,\epsilon) \sim \begin{cases} -x(-1) = x(1) & \text{for } t > 0 \\ x(-1) & \text{for } t < 0 \end{cases}$$

and there is a sharp *shock* or *transition layer* of nonuniform convergence in an order ϵ interval about $t = 0$.

Next suppose $x(1)+x(-1) > 0$, so $x(-1) > -x(1)$. The amplitude relation

$$A - x(-1) = (A + x(-1))\frac{(A - x(1))}{(A + x(1))}e^{-A/\epsilon}$$

is now solved iteratively to yield the positive solution

$$A \sim x(-1) + 2x(-1)\frac{(x(-1) - x(1))}{(x(-1) + x(1))}e^{-x(-1)/\epsilon} + \dots.$$

Since $e^{A\beta/2\epsilon} = \left(\frac{A+x(1)}{A-x(1)}\right) e^{A/2\epsilon}$, this implies the limiting solution

$$x(t,\epsilon) \sim x(-1)\left[\frac{(x(-1)-x(1)) - (x(-1)+x(1))e^{-x(-1)(1-t)/2\epsilon}}{(x(-1)-x(1)) + (x(-1)+x(1))e^{-x(-1)(1-t)/2\epsilon}}\right]. \tag{6.57}$$

Thus, $x(t,\epsilon) \to x(-1)$ for $-1 \le t < 1$, $x(1,\epsilon) = x(1)$, and there is a terminal layer of nonuniform convergence near the right endpoint. The behavior contrasts sharply, then, with the asymptotic solution with $x(1) < 0$ and $x(1) + x(-1) = 0$, where we found a sharp layer about the midpoint $t = 0$. Clearly, the dependence of the solution on $x(1) - x(-1)$ is not smooth near the ray $x(1) = -x(-1) < 0$.

For $x(1) < 0$ and $x(1) + x(-1) < 0$, we solve

$$A + x(1) = (A - x(1))\frac{(A + x(-1))}{(A - x(-1))}e^{-A/\epsilon}$$

to get $A = -x(1) - 2x(1)\left(\frac{x(1)-x(-1)}{x(1)+x(-1)}\right)e^{x(1)/\epsilon} + \dots$ and

$$x(t,\epsilon) \sim x(1)\left[\frac{(x(1) + x(-1)) - (x(1) - x(-1))e^{x(1)(t+1)/2\epsilon}}{(x(1) + x(-1)) + (x(1) - x(-1))e^{x(1)(t+1)/2\epsilon}}\right]. \tag{6.58}$$

Thus, the limiting solution is then $x(1)$, except in the boundary layer at the left endpoint $t = -1$.

For $x(1) > 0 > x(-1)$, we will need A to be purely imaginary, so we more conveniently rewrite the general solution as

$$x(t,\epsilon) = \epsilon\alpha \tan(\alpha(t - \epsilon B)) \tag{6.59}$$

for real constants α and B. The end conditions $x(\pm 1) = \epsilon\alpha\tan(\pm\alpha - \epsilon\alpha B)$ requires α to be nearly $\pi/2$, so we set $\alpha = \frac{\pi}{2} + \epsilon\gamma\alpha$ to obtain $x(\pm 1) \sim \frac{1}{B \mp \gamma}$. This uniquely determines the limiting values for γ and B as well as the limiting solution

$$x(t,\epsilon) \sim \left(\frac{\epsilon\pi x(1)x(-1)}{2x(1)x(-1) + \epsilon(x(-1) - x(1))}\right)$$
$$\times \tan\left(\frac{\pi[2x(1)x(-1)t - \epsilon(x(-1) + x(1))]}{2[2x(1)x(-1) + \epsilon(x(-1) - x(1))]}\right), \tag{6.60}$$

which tends to zero like ϵ within $[-1, 1]$ and has a nonzero limit only in an order ϵ neighborhood of the endpoints when the corresponding endvalue is nonzero.

In all cases, then, the limiting solution away from $t = 0$ or ± 1 is one of the constants 0 or $x(\pm 1)$. The precise asymptotic nature of the endpoint or midpoint layers of nonuniform convergence varies considerably, however. We note that Kevorkian and Cole (1996) and Lagerstrom (1988) consider the same two-point problem for the more complicated model equation $\epsilon\ddot{x} + x\dot{x} - x = 0$, where a closed-form solution is unavailable, but the asymptotic solution is similar to that for Burgers' equation.

Finally, we shall consider the second-order scalar autonomous equation

$$\epsilon^2\ddot{x} + f(x) = 0 \tag{6.61}$$

in conservation form, for t in the interval $-1 \le t \le 1$, and with $x(\pm 1)$ prescribed. (Note that the small parameter is here denoted by ϵ^2, rather than ϵ, for later convenience.) Rescaling the velocity and time, by setting

$$z = \epsilon \dot{x} \text{ and } \tau = t/\epsilon, \tag{6.62}$$

provides us with the first-order parameter-free system

$$\begin{cases} \frac{dx}{d\tau} = z \\ \frac{dz}{d\tau} = -f(x) \end{cases}$$

in the x-z phase plane. We note that rest points occur at $z = 0$ whenever $f(x)$ becomes zero, whereas

$$z\,dz + f(x)dx = 0$$

implies that orbits in the phase plane will satisfy

$$z = \pm\sqrt{2(E - V(x))}, \tag{6.63}$$

where

$$V(x) = \int^x f(s)ds$$

is the potential energy and E is the conserved total energy on any trajectory. The phase portrait is found by plotting allowed phase-plane trajectories for various E values. The boundary conditions imply that E and the sign of z must be determined so that

$$\epsilon \int_{x(-1)}^{x(1)} \frac{ds}{z(s)} = 2. \tag{6.64}$$

Thus, z must pass near zero along the solution's path, where V must stay below E. We indeed can anticipate that for most of the time t, x will remain near a maxima of V (so $\binom{x}{z}$ will be near a saddle point in the phase plane). That maximum will asymptotically determine the constant E.

Example 9 A very simple example of (6.61) is provided by the linear problem

$$\epsilon^2 \ddot{x} - x = 0, \quad x(-1) = 1, \quad x(1) = 2, \tag{6.65}$$

which has the unique solution

$$\begin{cases} x(t,\epsilon) = \frac{1}{(1-e^{-4/\epsilon})}\left[(1 - 2e^{-2/\epsilon})e^{-(t+1)/\epsilon} + (2 - e^{-2/\epsilon})e^{-(1-t)/\epsilon}\right] \\ z(t,\epsilon) = \frac{1}{(1-e^{-4/\epsilon})}\left[-(1 - 2e^{-2/\epsilon})e^{-(t+1)/\epsilon} + (2 - e^{-2/\epsilon})e^{-(1-t)/\epsilon}\right]. \end{cases}$$

Clearly,

$$\begin{cases} x(t,\epsilon) \sim e^{-(t+1)/\epsilon} + 2e^{-(1-t)/\epsilon} \\ \text{and} \\ z(t,\epsilon) \sim -e^{-(t+1)/\epsilon} + 2e^{-(1-t)/\epsilon} \end{cases}$$

is asymptotically negligible within $[-1, 1]$ and the solution features endpoint boundary layers with thicknesses of order ϵ. The corresponding phase-plane trajectory is given by

$$z(x) = \pm\sqrt{x^2 + 2E}$$

for $E = -4e^{-2/\epsilon}(1 - \frac{5}{2}e^{-2/\epsilon} + e^{-4/\epsilon})/(1 - e^{-4/\epsilon})^2$, so $x_{\min} \sim 2\sqrt{2}e^{-1/\epsilon}$ is asymptotically negligible and motion is fast for x away from zero, but slow near zero. The phase-plane pictures are given in Figures 20 and 21.

Example 10 Consider

$$\epsilon^2\ddot{x} + \sin\pi x = 0, \quad x(-1) = -2, \quad x(1) = 2. \tag{6.66}$$

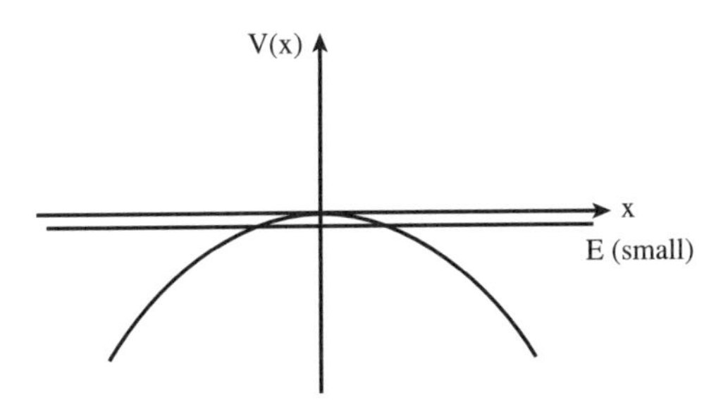

Figure 20. The potential $V(x) = -x^2/2$ for (6.65).

Each phase-plane trajectory must now satisfy

$$\frac{1}{2}z^2 - \frac{1}{\pi}\cos\pi x = E.$$

Because $-\frac{1}{\pi}\cos\pi x$ has the maximum value 1 at the odd integers, the phase-plane portrait determined in Figures 22 and 23 implies that the solution $x(t)$ of the two-point problem will correspond to some $E \sim \frac{1}{\pi}^+$ and $x(t)$ will increase monotonically. (For $E < \frac{1}{\pi}$, one cannot get from -2 to 2; the time required to travel along the orbit with $E = \frac{1}{\pi}$ is infinite.) Most of the time available will be spent near the two (unstable) rest points in $-2 < x < 2$, i.e., $x = \pm 1$ (since the rest points $x = 0$ and ± 2 are repulsive). Symmetry implies that the unique limiting solution $x(t)$ will be as in Figure 24. Construction of the boundary layers at $t = \pm 1$ and of the transition layer at $t = 0$ is no problem.

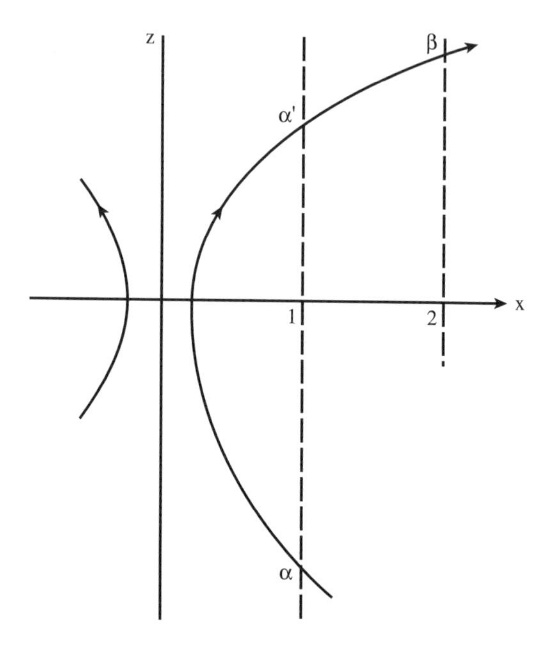

Figure 21. The solution of (6.65) for ϵ small.

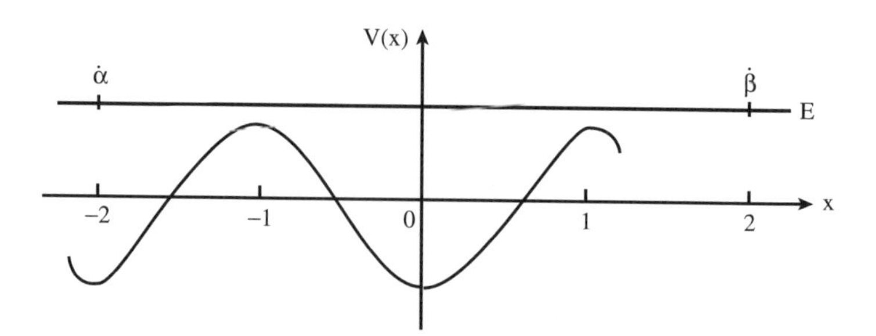

Figure 22. The potential $V(x) = -(\cos \pi x)/\pi$ for (6.66)

If we, instead, use boundary values $x(-1) = 2$ and any $x(1)$ between 1 and 3, the limiting solution within $[-1, 1]$ would be the same. Only the terminal jump would be different. If, however, we had $x(1) = 3$, the interior layer where the solution jumps from $x = -1^+$ to $x = 1^-$ would shift from $t \approx 0$ to nearly $t = 1/3$, i.e., two thirds of the "time" t would then be spent near $x = -1$. In this sense, the location of the interior jump is *supersensitive* to small changes in the value $x(1)$, near 1.

The solution would, likewise, be much different if we had the boundary values $x(-1) = -1/2$ and $x(1) = 3/4$. This requires an energy level $E \sim \frac{1}{\pi}^-$,

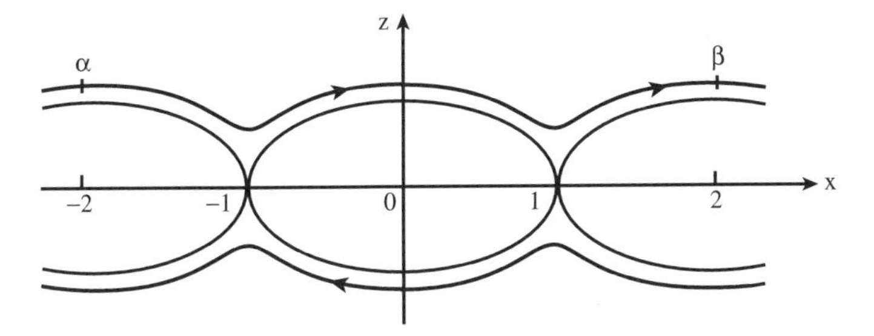

Figure 23. The phase plane solution of (6.66) for ϵ small.

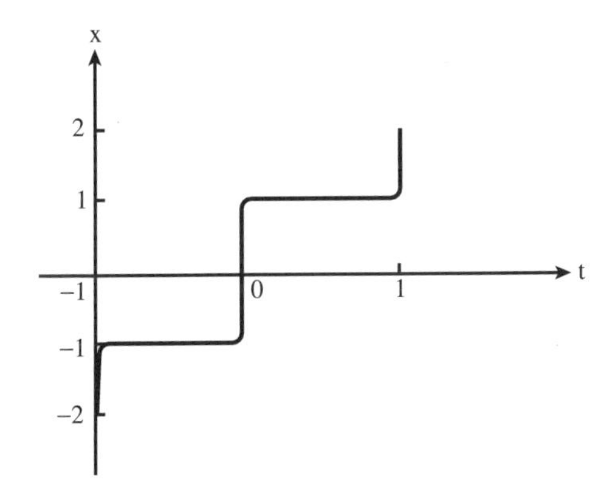

Figure 24. The solution $x(t, \epsilon)$ of (6.66) for ϵ small.

so the possibly periodic trajectory lies within the separatrix in the phase plane (cf. Figures 25 and 26).

The trajectory must start at a point α or α' on $x(-1) = -1/2$ and end at β or β' on $x(1) = 3/4$. It must pass near at least one of the rest points at $x = \pm 1$ in order to use up two units of time, but it can certainly go around the closed orbit any finite number of times. (The underlying asymptotic analysis will become more dubious as the number of cycles (and the path length) increases indefinitely.) Possibilities include the paths $\alpha'\alpha\beta$, $\alpha'\alpha\beta\beta'$, $\alpha\beta\beta'$, and $\alpha\beta\beta'\alpha'\alpha\beta$. The corresponding trajectories $x(t)$ will always have interior switchings, regularly spaced within $[-1, 1]$ (cf. Figures 27 to 29).

6.5 A combustion model exhibiting metastability

Reiss (1980) introduced the initial value problem

$$\dot{y} = y^2(1 - y), \quad t \geq 0, \quad y(0) \quad \text{given} \tag{6.67}$$

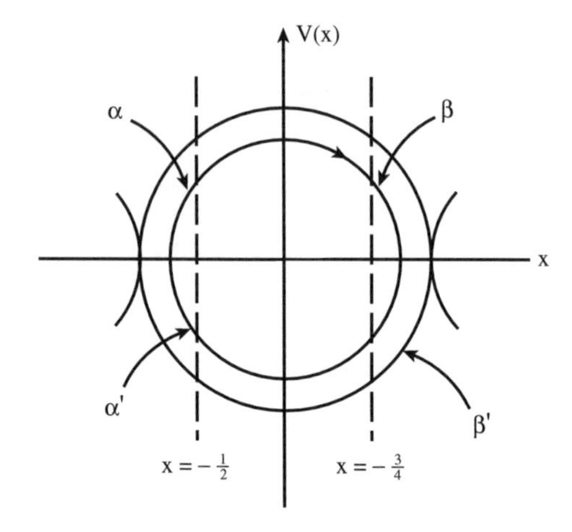

Figure 25. Phase plane portrait for (6.66) indicating possible solutions.

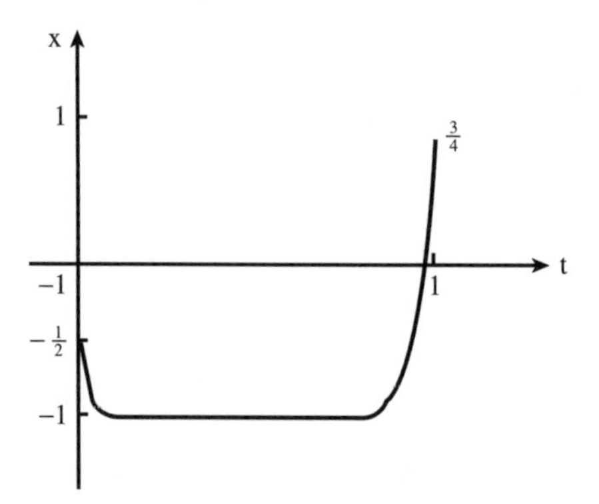

Figure 26. A solution of $\epsilon^2\ddot{x} + \sin \pi x = 0$ with two endpoint layers.

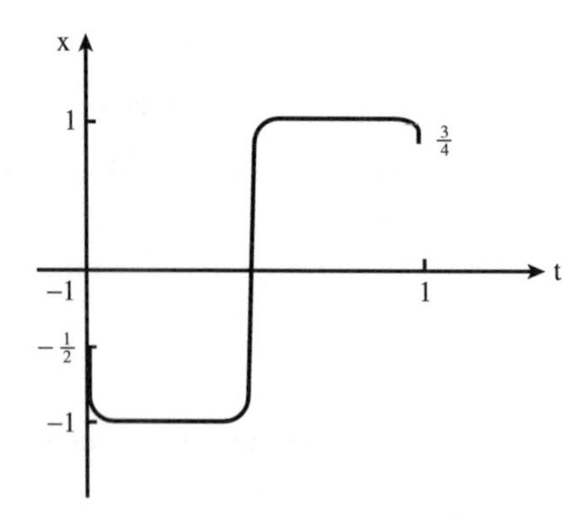

Figure 27. A solution of (6.66) with an interior layer.

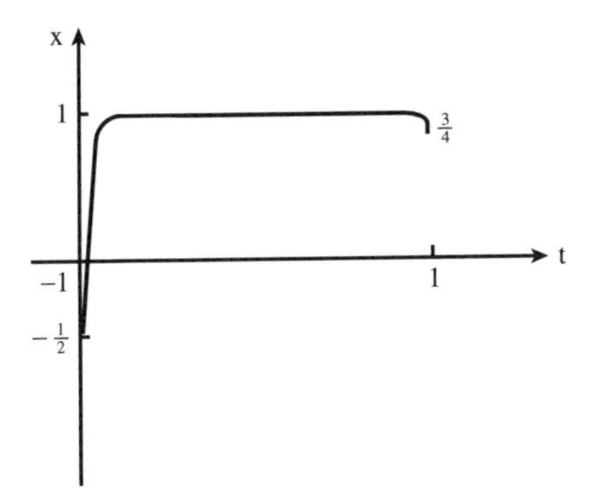

Figure 28. A solution of (6.66) with an endpoint layer.

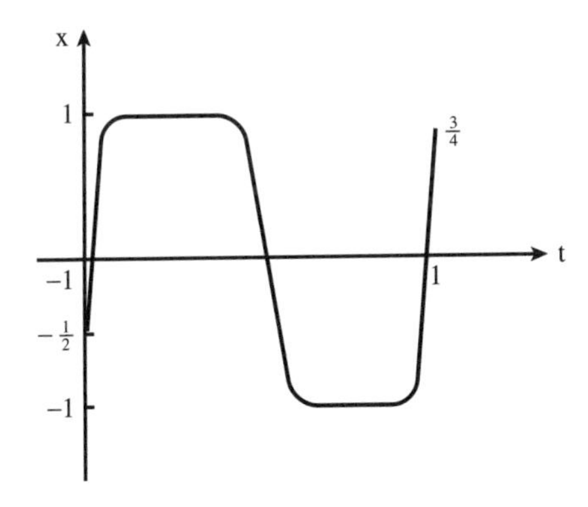

Figure 29. A solution of (6.66) with an interior and endpoint layers.

to model a jump phenomenon and it was subsequently developed as a combustion model by both Kassoy and Kapila. Then y represents the concentration of a reactant, $0 \leq y \leq 1$, and $y = 1$ represents the ignited state.

Separating variables, one obtains the monotonically increasing exact solution

$$\frac{1}{y} + \ln\left(\frac{1}{y} - 1\right) = -\kappa \equiv -t + t_0 \tag{6.68}$$

for $t_0 = \frac{1}{y(0)} + \ln\left(\frac{1}{y(0)} - 1\right)$, presuming $y(0) > 0$.

The solution $Y(\kappa)$ has the following tabular values

Y	κ
0.01	$-100 - \ln 99 \sim -104.60$
0.05	$-20 - \ln 19 \sim -22.94$
0.1	$-10 - \ln 9 \sim -12.20$
0.2	$-5 - \ln 4 \sim -6.39$
0.3	$-\frac{10}{3} - \ln\frac{7}{3} \sim -4.28$
0.4	$-\frac{5}{2} - \ln\frac{3}{2} \sim -2.91$
0.5	-2
0.6	$-\frac{5}{3} - \ln\frac{3}{2} \sim -1.26$
0.7	$-\frac{10}{7} - \ln\frac{7}{3} \sim -0.57$
0.8	$-\frac{5}{4} + \ln 4 \sim -0.14$
0.9	$-\frac{10}{9} + \ln 9 \sim 1.09$
0.95	$-\frac{20}{19} + \ln 19 \sim 1.89$
0.99	$-\frac{100}{99} + \ln 99 \sim 3.59$

Note, in particular, that the graph of $Y(\kappa)$ is quite lop-sided and that Y does not become zero until κ is a bit above 0.8. The approach to the stable asymptote 1 is quite rapid, while departure from the unstable asymptote 0 is relatively slow. Ninety eight percent of y's range occurs as κ varies from -105 to 3.6.

First, suppose the initial value is $y(0) = 1 - \epsilon$, so $t_0 = \frac{1}{1-\epsilon} + \ln\left(\frac{\epsilon}{1-\epsilon}\right)$ is large and negative as $\epsilon \to 0$. Then, one can conveniently seek the asymptotic solution as the Maclaurin expansion in ϵ of the form

$$y(t, \epsilon) = 1 + \epsilon y_1(t) + \epsilon^2 y_2(t) + \dots.$$

Substitution into the differential equation and initial condition requires successive terms to satisfy the initial value problems

$$\dot{y}_1 = -y_1, \quad y_1(0) = -1,$$
$$\dot{y}_2 = -y_2 - 2y_1^2, \quad y_2(0) = 0, \text{ etc.}$$

so we obtain the regular perturbation expansion

$$y(t, \epsilon) = 1 - \epsilon e^{-t} - 2\epsilon^2(e^{-t} - e^{-2t}) + \dots$$

which clearly displays the asymptotic stability of the ignited state 1 as $t \to \infty$.

Next, we will suppose $y(0) = \epsilon$, i.e. a very small perturbation above the trivial equilibrium. Note that the corresponding $t_0 = \frac{1}{\epsilon} + \ln\left(\frac{1}{\epsilon} - 1\right)$ is now large and positive. If we attempt a regular perturbation expansion, which must hold on bounded t intervals, we observe that

$$y(t, \epsilon) = \epsilon + \epsilon^2 t + \epsilon^3(t^2 - t) + \dots$$

doesn't lose its asymptotic validity until $t \to 1/\epsilon$, when all terms attain the same asymptotic size. This breakdown must be expected since all nontrivial solutions must ultimately increase to the stable rest state as t increases. This was convincingly demonstrated by the numerical results of Shampine and Reichelt (1996) as part of their effort to develop an Adams integrator for MATLAB5. For $\epsilon = 10^{-4}$, they show the rapid rise from the preignited state $y = 0$ to the ignited state $y = 1$ near $t = 10^4$ (cf. Figure 30). Less patient computing folks might have missed the excitement by stopping the integration too early. (If ϵ represents a seed, the corresponding pregnancy is not interminable; only $O(1/\epsilon)$ long!). The description, *metastability*, is therefore natural.

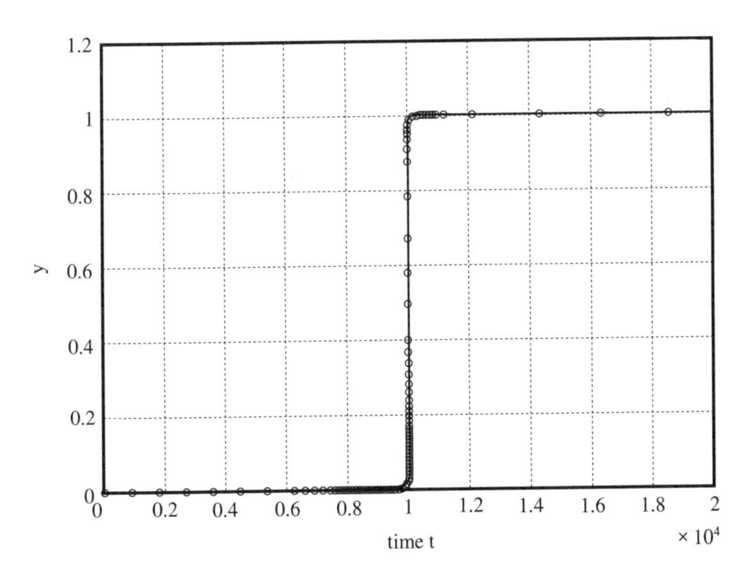

Figure 30. The computed solution to $\dot{y} = y^2(1 - y)$, $y(0) = 10^{-4}$.

To more fully comprehend the computation, let us introduce the compressed time

$$\sigma = \epsilon t$$

and rewrite the initial value problem in the traditional singular perturbation form

$$\epsilon \frac{dy}{d\sigma} = y^2(1-y), \quad y(0) = \epsilon, \sigma \geq 0.$$

It is then natural to seek regular perturbation solutions of the form

$$Y_L(\sigma, \epsilon) = \epsilon Y_{L1}(\sigma) + \epsilon^2 Y_{L2}(\sigma) + \ldots \text{ for } \sigma < 1$$

and

$$Y_R(\sigma, \epsilon) = 1 + \epsilon Y_{R1}(\sigma) + \epsilon^2 Y_{R2}(\sigma) + \ldots \text{ for } \sigma > 1,$$

anticipating the possibility of a rapid transition between these *outer* limits in some $O(\epsilon)$ interval about some unspecified value σ_0 near 1.

Note that the resulting outer limit

$$Y_L(\sigma, \epsilon) = \frac{\epsilon}{1-\sigma} - \epsilon^2 \frac{\sigma}{(1-\sigma)^2} + \ldots$$

becomes singular as $\sigma \to 1$, while the constant outer limit

$$Y_R(\sigma, \epsilon) = 1$$

results for $\sigma > 1$.

If wc introduce the stretched variable

$$\lambda = (\sigma - \sigma_0)/\epsilon,$$

we would naturally seek a transition layer solution

$$y(\sigma) = z(\lambda)$$

such that $z(\lambda)$ satisfies

$$\frac{dz}{d\lambda} = z^2(1-Z), -\infty < \lambda < \infty$$

such that

$$z(\lambda) \to \begin{cases} Y_L(\sigma, \epsilon) & \text{as } \sigma \to -\infty \\ Y_R(\sigma, \epsilon) & \text{as } \sigma \to +\infty. \end{cases}$$

By translation invariance, we can select

$$z(0) = Y(0).$$

Then, the uniqueness of solutions to initial value problems guarantees that we will obtain the transition layer profile

$$z(\lambda) = z\left(\frac{\sigma - \sigma_0}{\epsilon}\right) = Y(\kappa) = Y\left(\frac{\sigma}{\epsilon} - t_0\right)$$

But $t_0 = \frac{1}{\epsilon} + \ln\left(\frac{1}{\epsilon} - 1\right)$ then implies that $\sigma_0 \equiv \epsilon t_0$ is the location about which z passes through $Y(0)$. A careful examination of the numerical results shows that the transition is indeed well-described by a translation of the underlying exact solution $Y(\kappa)$. Similar asymptotic analyses can be successfully used for more difficult transition layer problems, where exact solutions are not available.

6.6 Exercises

1 Consider the scalar initial value problem

$$\epsilon \dot{x} = (t-1)x, \quad x(0) = 1.$$

(a) Find the exact solution $x(t, \epsilon)$.
(b) Show that the outer solution $X(t, \epsilon)$ for $0 < t < 1$ is asymptotically trivial.
(c) Determine $x(1, \epsilon)$ and its asymptotic size.
(d) Explain, on symmetry grounds, why $x(2, \epsilon) = 1$.
(e) Show that $x(t, \epsilon)$ becomes exponentially large for $t < 0$ and for $t > 2$.

2 Solve the scalar initial value problem

$$\epsilon \dot{x} = x - x^3$$

on $t \geq 0$, showing that the appropriate outer solution $X(t, \epsilon) \equiv -1, 0,$ or 1 is completely determined by the sign of $x(0)$.

3 Consider the enzyme kinetics problem

$$\begin{cases} \frac{dx}{dt} = -x + (x + \mu - \lambda)y, & x(0) = 1 \\ \epsilon \frac{dy}{dt} = x - (x + \mu)y, & y(0) = 0, \end{cases}$$

where λ and μ are given positive numbers and $\epsilon \to 0$.

(a) Show that the solution $\binom{X_0(t)}{Y_0(t)}$ of the reduced problem

$$\begin{cases} \frac{dX_0}{dt} = -X_0 + (X_0 + \mu - \lambda)Y_0, & X_0(1) = 1 \\ 0 = X_0 - (X_0 + \mu)Y_0 \end{cases}$$

will fail to satisfy $Y_0(0) = 0$.

(b) Show that the boundary layer correction term $\eta_0(\tau) = y(t) - Y_0(t)$ for $\tau = t/\epsilon$ must satisfy

$$\frac{d\eta_0}{d\tau} = -(1+\mu)\eta_0, \; \eta_0(0) = -Y_0(0) = -(1+\mu)^{-1}.$$

(c) Perform numerical computations for $\epsilon = 0.1$ and $\epsilon = 0.01$ and compare the results to the asymptotic approximations

$$x(t,\epsilon) \sim X_0(t)$$
$$y(t,\epsilon) \sim \frac{X_0(t)}{X_0(t)+\mu} - \left(\frac{1}{1+\mu}\right) e^{-(1+\mu)t/\epsilon},$$

where $\dot{X}_0 = -\lambda X_0/(X_0+\mu)$, $X_0(0) = 1$.

The latter solution should also be obtained computationally.

4 Determine the limiting solution to

$$\begin{cases} \dot{x} = yx \\ \epsilon\dot{y} = y - y^3 \end{cases}$$

on $t \geq 0$.

5 Solve the vector initial value problem for

$$\epsilon\dot{y} = (A_0 + \epsilon A_1)y$$

on $t \geq 0$, where A_0 is singular and

$$A_0 + \epsilon A_1 = \begin{pmatrix} 1-2\epsilon & 2-2\epsilon \\ -1+\epsilon & -2+\epsilon \end{pmatrix}.$$

Show that the limiting solution $Y_0(t)$ for $t > 0$ does satisfy the reduced problem $A_0 Y_0(t) = 0$.

6 (a) Determine the exact solution of the scalar Riccati equation

$$\epsilon\dot{y} = (1-t)y - y^2$$

on $t \geq 0$ with $y(0) > 0$.

(b) Integrate the exact solution by parts or use the ansatz of section 6.3 to determine the first three terms of the outer solution for $0 < t < 1$, noting its breakdown at $t = 1$.

(c) By performing a *very* careful numerical integration, show that the solution to the initial value problem tends to zero for $t > 1$. Thus, the limiting solution

$$y(t,\epsilon) \sim \begin{cases} y(0), \; t = 0 \\ 1-t, \; 0 < t < 1 \\ 0, \; t > 1 \end{cases}$$

features on initial layer of nonuniform convergence near $t = 0$ (unless $y(0) = 1$) and a derivative jump at $t = 1$.

7 Consider the two-point boundary value problem

$$\epsilon y'' - y' + (y')^3 = 0,\ y(0) = 0,\ y(1) = \frac{1}{2}.$$

(a) Solve the equation as a Riccati equation for $(y')^2$.

(b) Show that a limiting solution $Y_0(x)$ satisfies

$$Y_0(x) = \begin{cases} 0, & 0 \le x \le \frac{1}{2} \\ x - \frac{1}{2}, & \frac{1}{2} \le x \le 1. \end{cases}$$

(c) Find other angular limiting solutions.

8 (a) Find the exact solution of the two-point problem

$$\begin{cases} \epsilon^2 y'''' - y'' = x^2,\ 0 \le x \le 1 \\ y(0) = 0,\ y'(0) = 1,\ y'(1) = 2,\ y''(1) = 3. \end{cases}$$

(b) Determine the (related) boundary value problem satisfied by the limiting solution.

9 Determine the asymptotic solution to the two-point problem

$$\begin{cases} \epsilon^2 y'' = y - |x - \frac{1}{2}|,\ 0 \le x \le 1 \\ y(0) = y(1) = \frac{1}{2}. \end{cases}$$

In particular, determine $y'\left(\frac{1}{2}\right)$.

References on singular perturbations

1. J. D. Cole (1968), *Perturbation Methods in Applied Mathematics*, Blaisdell, Waltham, MA.
2. W. Eckhaus (1979), *Asymptotic Analysis of Singular Perturbations*, North-Holland, Amsterdam.
3. A. Erdélyi (1956), *Asymptotic Expansions*, Dover, New York.
4. S. Goldstein (1969), Fluid mechanics in the first half of this century, *Annual Review of Fluid Mechanics* 1, 1–28.
5. E. Hairer and G. Wanner (1991), *Solving Ordinary Differential Equations II. Stiff and Differential-Algebraic Problems*, Springer-Verlag, Berlin.
6. J. K. Kevorkian and J. D. Cole (1996), *Multiple Scales and Singular Perturbation Methods*, Springer-Verlag, New York.
7. P. A. Lagerstrom (1988), *Matched Asymptotic Expansions*, Springer-Verlag, New York.
8. J.-L. Lions (1973), *Perturbation Singuliéres dans les Problémes aux Limits et en Contrôle Optimal*, Springer-Verlag, Berlin.
9. J. D. Murray (1989), *Mathematical Biology*, Springer-Verlag, Berlin.

10. R. E. O'Malley, Jr. (1991), *Singular Perturbation Methods for Ordinary Differential Equations*, Springer-Verlag, New York.
11. E. L. Reiss (1980), A new asymptotic method for jump phenomena, *SIAM J. Appl. Math.* 39, 440–55.
12. L. A. Segel and M. Slemrod (1989), The quasi-steady state assumption: a case study in perturbation, *SIAM Review* 31, 446–77.
13. L. Shampine and M. Reichelt (1966), New ODE Solvers: Time and event location; when the ball bounces, when the fire flares, CODEE Newsletter, Winter, 13–15.
14. D. R. Smith (1985), *Singular-Perturbation Theory*, Cambridge University Press, Cambridge.
15. M. Van Dyke (1964, 1975), *Perturbation Methods in Fluid Dynamics*, Academic Press, New York; Annotated edition, Parabolic Press, Stanford, CA.
16. M. Van Dyke (1994), Nineteenth-century roots of the boundary-layer idea, *SIAM Review* 36, 415–24.
17. A. B. Vasil'eva, V. F. Butuzov, and L. V. Kalachev (1995), *The Boundary Function Method for Singular Perturbation Problems*, SIAM, Philadelphia, PA.
18. W. Wasow (1944), On the asymptotic solution of boundary value problems for ordinary differential equations containing a parameter, *J. Math. and Phys.* 23, 173–83.

General References

1. M. Abramowitz and I.A. Stegen (1965), eds., *Handbook of Mathematical Functions*, Dover, New York.
2. G. Birkhoff and G.-C. Rota (1978), *Ordinary Differential Equations*, Wiley, New York.
3. R. Bronson (1970), *Matrix Methods: An Introduction*, Academic Press, New York.
4. H.S. Carslaw and J.C. Jaeger (1959), *Conduction of Heat in Solids*, Clarendon, Oxford.
5. W.A. Coppel (1965), *Stability and Asymptotic Behavior of Differential Equations*, Heath, Boston
6. N.D. Fowkes and J.J. Mahony, *An Introduction to Mathematical Modelling*, Wiley, Chichester.
7. R. Haberman (1987), *Elementary Applied Partial Differential Equations*, Prentice-Hall, Englewood Cliffs, NJ.
8. W. Hahn (1963), *Theory and Application of Liapunov's Direct Method*, Prentice-Hall, Englewood Cliffs, NJ.
9. E. Kamke (1944), Differentialgleichungen Lösungs methoden und Lösungen, Becker & Erler, Leipzig.
10. T.W. Körner (1988), *Fourier Analysis*, Cambridge University Press, Cambridge.
11. A.C. Lazer and P.J. McKenna (1990), Large-amplitude periodic oscillations in suspension bridges: Some new connections with nonlinear analysis, *SIAM Review* 32, 537–78.
12. R.M.M. Mattheij and J. Molenaar (1996), *Oridinary Differential Equations in Theory and Practice*, Wiley, Chichester.
13. C.B. Moler and C.F. Van Loan (1978), Nineteen dubious ways to compute the exponential of a matrix, *SIAM Review* 20, 801–36.
14. F.W.J. Olver (1974), *Introduction to Asymptotics and Special Functions*, Academic Press, New York.
15. H. Poincaré (1892), *Les Méthodes Nouvelles de la Mechanique Céleste*, Gauthier-Villars, Paris.
16. J.C. Polking (1995), *ODEs using MATLAB*, Prentice-Hall, Englewood Cliffs.
17. D.A. Sanchez (1968), *Ordinary Differential Equations and Stability Theory: An Introduction*, Freeman, San Francisco.
18. L.A. Segel (1977), *Mathematics Applied to Continuum Mechanics*, Macmillan, New York.
19. D. Zwillinger (1989), *Handbook of Differential Equations*, Academic Press, Boston.

Index